THE SPIRIT OF GREEN

The Spirit of Green

The Economics of Collisions and Contagions in a Crowded World

William D. Nordhaus

PRINCETON UNIVERSITY PRESS

PRINCETON AND OXFORD

Published by Princeton University Press
41 William Street, Princeton, New Jersey 08540
6 Oxford Street, Woodstock, Oxfordshire OX20 1TR

press.princeton.edu

Library of Congress Cataloging-in-Publication Data

Names: Nordhaus, William D., author.
Title: The spirit of green : the economics of collisions and contagions in
    a crowded world / William D. Nordhaus.
Description: 1st Edition. | Princeton : Princeton University Press, 2021. |
    Includes references and index.
Identifiers: LCCN 2020054805 (print) | LCCN 2020054806 (ebook) |
    ISBN 9780691214344 (hardback) | ISBN 9780691215396 (ebook)
Subjects: LCSH: Economic development—Environmental aspects. |
    Climatic changes—Economic aspects. | Greenhouse gas mitigation—Economic
    aspects. | Industrial productivity. | Climatic changes—Social aspects.
Classification: LCC HD75.6 .N657 2021 (print) | LCC HD75.6 (ebook) |
    DDC 338.9/27—dc23
LC record available at https://lccn.loc.gov/2020054805
LC ebook record available at https://lccn.loc.gov/2020054806

British Library Cataloging-in-Publication Data is available

Editorial: Joe Jackson, Josh Drake
Jacket Design: Derek Thornton
Production: Erin Suydam
Publicity: James Schneider, Kate Farquhar-Thomson
Copyeditor: Wendy Lawrence

Jacket image: Amager Bakke, Copenhagen, Denmark.
Photo: Gonzalez Photo / Astrid Maria Rasmussen / Alamy

This book has been composed in Adobe Text Pro.

Printed on acid-free paper. ∞

Printed in the United States of America

10 9 8 7 6 5 4 3 2 1

TABLE OF CONTENTS

1

Preface

Growing up in the high desert of New Mexico, I saw green as a welcome relief from the arid landscape. "It is so green," my father would say as we drove up to the family cabin in the mountains. That usually meant he hoped there was enough water in the stream for trout fishing. Green to my father meant trout in the pan.

My view of the meaning of "Green" has changed since my willowed fishing days. Green has taken on a life of its own, becoming a social movement that reflects a new approach to individual actions, companies, political activities, and laws. It is an interconnected set of ideas about the dangerous side effects of modern industrial societies and how we can cure, or at least curb, them. In this book, "Green" with a capital *G* represents the movement to deal with the collisions and contagions of the contemporary world. When written with a lowercase *g*, "green" refers to the perceived color of trees and plants.

When I sketched this book in my mind a decade ago, I hoped to address the challenges raised by economic growth and globalization and their unintended side effects. The side effect that has engaged me most is climate change, and the search for policies to slow global warming generated many of the ideas in this book. As the final words of this book are being written, the world is presently haunted by another scourge, the pandemic caused by the novel coronavirus.

Plagues are as old as climate change is new, but the solutions have a common core of approaches. Societies need to combine the ingenuity of private markets with the fiscal and regulatory powers of governments. Private markets are necessary to provide ample supplies of goods such as food and shelter, while only governments can provide collective goods such as pollution control, public health, and personal safety. Operating the well-managed society without both private markets and collective actions is like trying to clap with one hand. This book discusses how to harness the strengths of both private and public forms of social organization to find effective solutions to the complex challenges faced by interrelated industrial societies.

The impact of the environmental, or Green, movement is examined in various areas here. While most people think of pollution as the major spillover of modern life, the world has learned that pandemics can be deadly by-products of everyday personal and economic transactions. Green means not only a clean planet but also a world free of devastating infectious diseases like COVID-19.

Blueprint for a Green Planet

The chapters of this book cover a wide array of social, economic, and political questions that are examined from a Green vantage point. They include established areas such as pollution control, reduction of congestion, and global warming. But they also involve new frontiers such as Green chemistry, taxes, ethics, and finance.

We begin our journey with the cover of this book, which features a futuristic piece of architecture called "Copenhill," recently completed in Copenhagen, Denmark. This building combines interior offices with a trash-to-electricity plant, a hiking trail, and a chairlift serving grassy beginner-to-expert ski slopes. Few people would imagine Copenhill as the icon of the Green age because of its association with garbage, but it shows how different components of our lifestyles—from production to working to skiing—can be innovatively integrated.

Copenhill is a monument to Green architecture, which is usefully described by one of its advocates, James Wines, as follows:

"Green architecture is a philosophy of architecture that advocates sustainable energy sources, the conservation of energy, the reuse and safety of building materials, and the siting of a building with consideration of its impact on the environment." *Sustainability* is the key here. In Green architecture it means minimizing the harmful environmental impact of buildings through efficient design and the use of renewable resources. More generally, in a theme running throughout this book, a sustainable society is one that operates to ensure that future generations can enjoy living standards at least as ample as those of today.

The built environment is the most durable tangible feature of human civilization. Aside from a few tools, the oldest human artifacts are buildings. These include Egyptian pyramids, Roman aqueducts, Indian pueblos, and Gothic cathedrals. Most structures last at least a half century, compared to a decade for cars or a couple of years for smartphones. Because buildings are so prominent and last so long, they are a useful illustration of the importance of the application of Green principles.

While the Spirit of Green is useful as a blueprint for structures and other tangible goods, it is even more influential as a conceptual framework for the design of institutions, laws, and ethics for an interconnected society. The analytical foundations of Western economies are built on the ideas of Adam Smith and the nineteenth-century liberals. Their approach emphasizes competitive markets free of monopoly and fraud. Economic insights of an earlier age remain a critical component of a prosperous society, but they must increasingly be balanced with the philosophy required to correct market and nonmarket flaws.

This book describes Green philosophy and its application to a globalized and technologically sophisticated society. In some cases, as in the building on the cover of this book or in new vehicles or chemicals, the approaches are literally or figuratively concrete.

However, some of the most important Green approaches are organizational or institutional or attitudinal. Changing our tax system, developing more accurate measures of national output, improving the incentives for green energy, using market instruments to reduce pollution, and improving the ethical norms for individuals and firms—these are ways of altering society that require no steel or concrete but rather changes in attitudes and laws.

Before turning to the different themes that follow, I must give a nod of thanks to the friends and colleagues who have taught me so much. I particularly salute my teachers from an earlier generation: Tjalling Koopmans, Paul Samuelson, Robert Solow, and James Tobin.

Additionally, I give thanks to contributors to the invisible college of environmental and economic thinking. They include George Akerlof, Jesse Ausubel, Lint Barrage, Scott Barrett, William Brainard, Nicholas Christakis, Maureen Cropper, Dan Esty, Alan Gerber, Ken Gillingham, Geoffrey Heal, Robert Keohane, Charles Kolstad, Matt Kotchen, Tom Lovejoy, Robert Mendelsohn, Nick Muller, Nebojsa Nakicenovic, John Reilly, Jeffrey Sachs, Cass Sunstein, David Swenson, Martin Weitzman, Zili Yang, and Gary Yohe.

The last salute goes to my brother Bob, an inspiration in life and the law, who devoted his talents to writing Green ideals into federal energy and environmental legislation.

All remaining errors and impractical flights of fancy belong to the author.

———

I write the final words of this book on January 21, 2021, the day after Joseph Biden became the 46th President of the United States and the world left the dark ages of the Trump years. The new administration, along with governments and citizens around the world, face challenges, Green and beyond, more daunting than at any time in half a century. But good will, sound science, and the rule of democratic institutions will serve as beacons to light our way over the coming years.

PART I

Foundations of a Green Society

Green History

The Green movement reviewed here starts near my home in New Haven, Connecticut, with a forester, Gifford Pinchot. He endowed the Yale Forestry School, wielded the ax in clear-cutting forests, and pioneered our country's early forest policies. The review ends in the same place with a talented environmental lawyer at the same school, now Yale's School of the Environment. We will see how the movement has been transformed when we review Professor Dan Esty and his collection of radical ideas to protect and preserve our planet.

Pinchot, Muir, and the Founding of American Environmentalism

Environmentalism, as we know it today, was born in the late 1800s. For almost a century, it concentrated on the management and preservation of natural resources, particularly forests and wilderness areas. Natural resources provide a mixture of market and nonmarket services, and many of the most contentious debates in the early years related to the relative importance of relying on the market versus government. The two founders of environmental thinking, Gifford Pinchot and John Muir, provided the basis of the later debates.

The history of American environmentalism began with Gifford Pinchot. He is a familiar name at Yale, where he graduated in 1889 and later endowed Yale's Forestry School. He came from a wealthy family of lumber magnates, who had a practice of clear-cutting vast swaths of western forests for their operations. Some of his thinking, such as social views on eugenics and environmental views on clear-cutting, are now largely discredited, but he was a pioneer in forestry science.

Pinchot believed that forests were essential national assets as sources of timber, but he also thought that private firms mismanaged forest resources. The primary failure of firms was too short a time horizon (or too-high discount rates, in modern parlance). He wrote, "The forest is threatened by many enemies, of which fire and reckless lumbering are the worst." The role of government, in his view, was to ensure the proper use of forest assets, protecting forests from their enemies.

Pinchot was among the first proponents of *sustainability*, a core principle of the Green movement. He wrote:[1]

> The fundamental idea in forestry is that of perpetuation by wise use—that is, of making the forest yield the best service possible at present in a way that its usefulness in the future will not be diminished, but rather increased.

This statement puts into words one of the deepest ideas of modern environmental economics. *Sustainable consumption* (whether from timber harvesting or the economy, more generally) is the amount that can be consumed while leaving the future as well off as today.

Pinchot was not only a visionary but also a practitioner. While he thought forests were valuable for multiple uses, he primarily emphasized the harvesting of timber, which he saw as "a regular supply of trees ripe for the ax." He emphasized that "many of the most serious dangers to the forest are of human origin. Such are destructive lumbering and excessive taxation on forest lands. . . . So high are these taxes . . . that [loggers] are forced to cut or sell their timber in haste and without regard to the future." His mission was to correct destructive practices in order to establish "practical forestry," which would make "the forest render its best service to man in such a way as to increase rather than to diminish its usefulness in the future."

The other iconic figure of that age was John Muir. If Pinchot was a man of the ax, Muir was a man of the boot. Born in Scotland, he immigrated to Wisconsin at the age of 11, worked odd jobs, farmed, had suffered through a short university career, and then discovered his love of walking and nature. Muir, a major contributor to the establishment of America's National Park System, founded the Sierra Club and the "preservationist" wing of modern environmentalism.

In his twenties he began a lifelong pattern of walking around the country. He undertook a 1,000-mile walk across the country. When he reached the ocean at the Florida Keys, his romantic spirit caught fire:[2]

> Memories may escape the action of will, may sleep a long time, but when stirred by the right influence, though that influence be light as a shadow, they flash into full stature and life with everything in place. . . . I beheld the Gulf of Mexico stretching away unbounded, except by the sky. What dreams and speculative matter for thought arose as I stood on the strand, gazing out on the burnished, treeless plain!

When he later founded the Sierra Club, he put these feelings into its charter, which held that its purpose was "to explore, enjoy, and render accessible the mountain regions of the Pacific Coast" and to enlist support "in preserving the forests and other natural features of the Sierra Nevada Mountains." Since then, the Sierra Club has broadened its mission to "explore, enjoy and protect the wild places of the Earth [and] to practice and promote the responsible use of the Earth's ecosystems and resources."

Human-Centrism versus Biocentrism

One theme in Muir's writing is the human-centered idea that valuable natural sites should be protected and preserved for future generations (this is called the human-centric, or *anthropocentric*, approach). Human values today form the basis for virtually all legal and economic analyses of the value of natural resources.

A second and distinct theme is an ecological view that nature has its own value independent of humans and should therefore be preserved even if no humans can enjoy it (this is the *biocentric* approach).[3]

Most people have an intuitive sense that nature is intrinsically valuable even if they do not know how to value it or how to make the trade-off between human and nonhuman concerns. One example of the biocentric approach is the animal rights movement, which holds that animals have rights or interests independent of those of humans.

From an economic perspective, we might ask, "What is the value of a forest or an ecosystem?" More generally, what is the value of natural systems? It is useful to distinguish three different ways of valuing the environment, as shown in figure 2-1. Pinchot and many market-oriented people emphasize the importance of circle *A*, which is the market value of products such as lumber. We should not downplay the importance of market outputs. People always need food, shelter, and clothing and in the modern era enjoy their cell phones, television shows, and concerts.

In addition to the market values in *A*, however, we must also recognize the nonmarket activities in circle *B*. These include leisure and family life, as well as services of natural assets such as a walk on the beach or a hike in the mountains. Perhaps circle *B*'s nonmarket services of value to humans are just as important as circle *A*'s market activities. From a conceptual point of view, both *A* and *B* are anthropocentric (human centered) in the philosophical sense of promoting the welfare of humans, but these serve human goals through different mechanisms.

Circle *C* adds a new dimension to values by suggesting that nonhuman species or ecosystems or individual animals have an intrinsic value *independent of their value to humans*. This critical point is worth further exploration. Most social sciences such as economics, as well as legal theories, include only the preferences or welfare of humans in society's goals and preferences.

However, some philosophers and environmentalists (as well as animal rights groups) would like to extend the boundary of interests and values to include the welfare of nonhuman species.[4] In

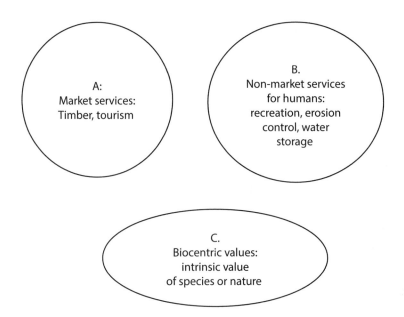

FIGURE 2-1. Alternative value systems

Circle *A* represents the market values of a forest, which is maximized under the Pinchot approach. Circle *B* contains the nonmarket values of the system, which will not be efficiently provided by an unregulated market. Circle *C* contains items that are not necessarily valued by humans but have their intrinsic worth.

environmental studies this approach is sometimes called biocentrism or deep ecology. Here is the way a proponent, philosopher Paul Taylor, describes the fundamental principles behind biocentrism:

> Our duties toward the Earth's nonhuman forms of life are grounded on their status as entities possessing inherent worth. They have a kind of value that belongs to them by their very nature, and it is this value that makes it wrong to treat them as if they existed as mere means to human ends. It is for their sake that their good should be promoted or protected. Just as humans should be treated with respect, so should they.[5]

Taylor's approach contrasts with (or some would say supplements) the standard analysis in law and economics in which actions should be taken to improve the welfare or preferences of humans. Note that asserting an intrinsic value of nonhuman life is different from saying that humans value nonhuman life. Most people would agree that preserving polar bears or coral reefs is a valuable activity because humans love them. They might add that these valuable life forms have intrinsic value. More difficult cases for those holding to the intrinsic value of life would be mosquitoes or jellyfish, which many humans would like to kill off but biocentrists might protest to be valuable in themselves.

Return to our discussion of Pinchot and Muir. Pinchot clearly focuses primarily on the values of circle A but in doing so insists that government regulation is necessary to ensure that the values in circle A are optimized. Muir's view is broader. He clearly believes that the nonmarket values in circle B are important, but sometimes he argues for protecting nature for its own intrinsic value (circle C).

It is likely that while Muir possessed some of both the human-centered and the biocentric spirit, he did not distinguish them as sharply as we might today. His biocentric view comes through clearly in his defense of alligators, where he wrote, "Many good people believe that alligators were created by the Devil, thus accounting for their all-consuming appetite and ugliness. . . . From the same material he has made every other creature, however noxious and insignificant to us. They are earth-born companions and our fellow mortals."[6] At the same time, he was practical and recognized the importance of mobilizing people who were interested in nature as an uplifting experience. Alligators have no votes and few sympathizers.

The Tragedy of the Commons

One of the most influential articles in all of the environmental sciences is Garrett Hardin's "The Tragedy of the Commons," published in 1968.[7] Trained as a microbiologist, Hardin quickly turned to a career of public advocacy, critiquing population and economic growth. He represented what has become an antimarket theme in modern environmentalism.

The basic thesis in Hardin's tragedy was that the competition of Adam Smith's unregulated market or "invisible hand" (discussed at length in chapter 4) can lead to ecological and human disaster. Hardin argued that Smith's analysis "contributed to a dominant tendency of thought that has ever since interfered with positive action based on rational analysis, namely, the tendency to assume that decisions reached individually will, in fact, be the best decisions for an entire society."[8]

Hardin provided many examples of the inefficiency of market forces, but he focused on the explosive growth of the human population. Many people were advocating technical solutions such as farming the seas or developing new hybrid grains. He argued that these were fruitless: "No technical solution can rescue us from the misery of overpopulation."[9]

He reasoned that a couple who adds another person to the family is like the herdsman who adds another animal to his herd and thereby contributes to the overgrazing of the commons:

> Each man is locked into a system that compels him to increase his herd without limit—in a world that is limited. Ruin is the destination toward which all men rush, each pursuing his own best interest in a society that believes in the freedom of the commons. Freedom in a commons brings ruin to all.[10]

The tragedy of the commons is today viewed as an example of economic inefficiencies caused by externalities (more specifically, common-property resources, discussed in detail later). Overgrazing occurs when vegetation is eaten so intensively that it does not have time to regenerate. The individual herdsman does not pay for the loss of regenerative capacity, and a fertile pasture thereby turns into an arid scrubland. This syndrome is also seen where common-property resources, such as the oceans or the air, are degraded because their exploitation is underpriced.

Rachel Carson's Pioneering Contribution

When environmental theories sprouted in the late 19th century, they attracted limited attention. Political struggles in that era of American capitalism focused on the tariff question, gold and silver, labor battling capital, and the rise of monopolies and the struggle to contain them, with periodic wars and depressions.

After World War II, the scale of economic activity began to put increasing pressures on the land, air, and water. One of the central figures in alerting the public and the political leaders to environmental concerns was scientist-poet Rachel Carson (1907–1964).

Carson was born in a small town north of Pittsburgh and pursued studies in marine biology. She was fascinated by the ocean and began writing radio programs and articles. Her writings described the seas in eloquent passages: "Who has known the ocean? Neither you nor I, with our earth-bound senses, know the foam and surge of the tide that beats over the crab hiding under the seaweed of his tide-pool home."[11]

Her work in conservation biology led her to become concerned about the widespread use of pesticides. The most important and damaging was DDT, used to control everything from head lice in soldiers to mosquitoes in the tropics. Based on her research, she published a warning book, *Silent Spring* (1962), which described the dilemmas that societies faced in attacking nuisances with chemicals:[12]

> No responsible person contends that insect-borne disease should be ignored. The question that has now urgently presented itself is whether it is either wise or responsible to attack the problem by methods that are rapidly making it worse. The world has heard much of the triumphant war against disease through the control of insect vectors of infection, but it has heard little of the other side of the story—the defeats, the short-lived triumphs that now strongly support the alarming view that the insect enemy has been made actually stronger by our efforts. Even worse, we may have destroyed our very means of fighting.[13]

The book was widely acclaimed by environmentalists and scientists. It caught the attention of the advisers of President John F. Kennedy and then of the president himself even before it was published. Kennedy gave *Silent Spring* a public endorsement and then ordered the President's Science Advisory Committee to study various health and environmental questions about pesticide use. The publicity lent momentum to the Kennedy administration's proposals for environmental legislation.[14]

However, Carson's critique opened a new chapter in environmental politics by provoking a ferocious response from the affected firms. Companies threatened to sue the publisher to prevent the publication of *Silent Spring*, and chemical companies, such as Velsicol, undertook opposition research to combat the damage to their reputations and bottom lines. This was not the first time the chemical-industrial complex had attacked environmental critics, but it was one of the most aggressive and set the stage for similar struggles between scientists and companies in areas such as tobacco, acid rain, and global warming.

Radical Ideas for Saving the Planet

This chapter ends with a bow to the recognition of the importance of radical new ideas for societal improvement. We will see again and again throughout this book how new technologies and ideas caused problems that other ideas and technologies helped to solve. Economic improvements led to the explosive growth of human populations in cities, which required armies of horses for transportation, which in turn left mountains of horse manure in their tracks. The mountains disappeared only after the newly invented automobile—often despised by modern-day environmentalists—replaced horses and cleaned the city streets.

Rolling forward to today, we face similar challenges, as will be shown in the chapters that follow. These range from local issues like congestion to global issues such as climate change. The theme that runs throughout this book is that humanity's problems are solvable if we listen carefully and critically to radical ideas, new and old.

One place to look is the recent book edited by Dan Esty, *A Better Planet: Forty Big Ideas for a Sustainable Future*.[15] Esty has a career that spans private research and public advocacy. He is a professor at Yale Law School and the Yale School of the Environment; a former commissioner of the Connecticut Department of Energy and Environmental Protection; and a prolific author and advocate for environmental improvement through innovation.

A Better Planet has forty chapters, each of which takes an environmental issue and proposes a radical solution. One example is Tracy Mehan's proposal for rethinking the concept of *wastewater*.[16] Water is scarce in many parts of the world. Yet, the American West would have vast supplies of water if we would reuse rather than waste the water. Using new technologies, what goes down the drain can be treated and sent back through our taps. Neither droughts nor dwindling snowpacks can reduce the supply of waste-to-tap water.

———

We have launched our journey with a short history of the evolution of the Green movement by describing some of its leaders. This history emphasizes that radical ideas and technologies, rather than axes and soldiers, will solve our environmental problems. J. M. Keynes emphasized this point when he introduced his radical new ideas into economics:[17]

> The ideas of economists and political philosophers, both when they are right and when they are wrong, are more powerful than is commonly understood. Indeed, the world is ruled by little else.

The insights of leaders such as Pinchot, Muir, Hardin, Carson, and Esty have rippled through society and modern environmental policy, affecting deeply views of how we should govern society and the natural world. We now turn to see how far those ripples have traveled.

Principles of a Green Society

On a first encounter with the Green movement, I had little apprecia-
tion of how deeply it had penetrated modern thinking. Newspapers
write about pandemics and climate change, but works on other areas
such as Green ethics, finance, taxation, and corporate planning were
not on my bookshelf.

All these topics—from rules for individuals to the challenge of
global pollution and pandemics—are part of the sweeping landscape
of the Green movement. But what is the architecture of Green? How
do Green principles fit into the conception of a well-managed society? What are the key tenets of Green thinking? We begin with these
questions to position our thinking for the different areas.

The Goal of a Well-Managed Society

Before analyzing different fields of the spirit of Green, it will be helpful to embed the discussion in a more general philosophical view of
the nature of the kind of society that we desire.

The ideal society in my mind involves a structure of institutions,
attitudes, and techniques for promoting a just and prosperous nation.
For simplicity, I call it a *well-managed society*. This topic has occupied
political and economic philosophers for centuries. While this

synthesis is my own, it draws upon a long line of political and economic thinkers such as John Stuart Mill, Arthur Pigou, Robert Dahl, Paul Samuelson, and John Rawls. The ideas extend far beyond the scope of this book, and the purpose here is to sketch the elements of a well-managed society that are shared with the spirit of Green.[1]

A close relative of the ideas described here is what Harvard philosopher John Rawls called a "well-ordered society." In Rawls's words, a well-ordered society is the result of "bringing together certain general features of any society that it seems one would, on due reflection, wish to live in and want to shape our interests and character."[2]

I use a different name—a well-managed society—to distinguish the concept from Rawls's and because it has a different emphasis. Rawls concentrated on justice in his writings on the good society. The emphasis here, as in much economic writing, adds the additional goal of efficiency to that of justice.

Four Pillars

When considering Green goals, a well-managed society has four pillars. First, it requires *a body of laws* that defines the relations among people. The laws should enforce civil conduct and civil rights, define and enforce property rights and contracts, and promote equality and democracy. Good laws allow people to interact in ways that ensure reliable transactions along with the fair and efficient adjudication of disputes.

The second pillar is a set of well-developed *markets for private goods*, which are goods for which firms and consumers pay the full costs of their provision and enjoyment. The key mechanism for the effective provision of private goods is through supply and demand on markets. It is here that individuals and firms, pursuing their own interests in trade and exchange, promote efficiency through Adam Smith's invisible-hand mechanism.

Third, society must find the techniques to deal with *public goods*, or *externalities*. These are activities whose costs or benefits spill outside the market and are not captured in market prices. They include negative spillovers like pollution and infections as well as positive spillovers like new knowledge. A well-managed society will ensure that major negative externalities are corrected through governmental laws that promote negotiations and liability for damage through powers such as regulations and taxes. Moreover, in areas where governmental actions are missing or incomplete, it will be necessary for individuals and private institutions to be mindful of their external impacts.

Finally, a well-managed society requires that governments pursue equality in institutions along with corrective taxation and expenditures to ensure that the distribution of economic and political opportunities and outcomes are equal and just. This goal has become particularly important with the growing economic disparities over the last half century. To take just one example, the wealth of the top 1% of families was 15 times the average wealth in 1963. That grew to 50 times the average wealth by 2016. It is important that noxious externalities should not pile on to existing inequalities by adding more.

Of course, to state the requirements of a well-managed society does not provide clear answers as to how to implement these goals. Reducing inequality is often divisive because the haves do not always part with what they have without resistance. Moreover, nations cannot and should not regulate every minor externality, such as a messy yard or burping in public. But the general principles are clear and have weighty implications for political decisions, as well as for the ethics of private participants such as corporations or individuals.

Pillars for a Green Society

The goals of a Green society are nested inside the goals of a well-managed society, with emphasis on particular harms and remedies. The first pillar of the legal structure makes individuals and other entities answerable for their actions. For example, it would insist that people are responsible for their damages when they drive an automobile and that reckless behavior is appropriately penalized. The second pillar of the market is a guide for people's market behavior—both as suppliers as well as consumers—using the signals of prices, wages, and incomes. A properly functioning market vastly simplifies life, as it provides a cornucopia of goods and services through domestic and international trade.

The third principle—dealing with important public goods and focusing especially on harmful externalities—is the heart of the spirit of Green. This principle covers a vast spectrum of spillovers from local litter to global warming, dealing with issues visible and invisible, transient and long-lived, irritating and deadly. We will see examples of harmful externalities from many sources throughout this book.

The final principle reminds us that people have vastly different opportunities and outcomes. It is easy to become entranced with finding the most efficient and effective tools and outcomes, but we must be attentive to their distributional impacts, particularly on low-income people and nations. Environmental justice is part of the more general goal of justice and equality.

Themes of the Spirit of Green

Most of the readers of this book will be familiar with Green thinking in their own areas of interest, but they may be surprised by how these ideas have spread to other areas. Although the strands of Green ideas may appear unconnected, a few core concepts will emerge throughout the discussion. These involve the impact of globalization, the increasing prevalence of collisions and contagions, the importance of federalism, the basic prescriptions for policies, and the mechanisms for action.

THE IMPACT OF GROWTH AND GLOBALIZATION

Why does Green matter? The Green movement is a response to an increasingly crowded world. It reflects the many ways that—in a globalized, rapidly changing, interconnected, and technologically advanced world—we are colliding with each other all the time. In an earlier era, plagues would spread slowly by horse or by ship and take months to cross the globe, often burning out along the way. In the modern era, a flight from China to Europe or the United States can spread deadly pathogens overnight—indeed, before scientists can even identify them.

Sometimes, our interactions are innocuous, such as when we brush against each other on the sidewalk. More consequential are the frequent collisions on the roadways, as well as the infrequent but frightening ones in the airways. Most damaging to a society are the interactions that occur when firms emit pollution that kills people or changes the climate; when companies move production overseas and displace workers and hurt communities; when companies knowingly produce dangerous products that sicken, maim, or kill people; and when pandemics spread around the world, killing hundreds of thousands and sending economies into a tailspin. To give these tendencies a colorful name, I will label as "Brown" these forces of degradation.

COLLISIONS, CONTAGIONS, AND EXTERNALITIES

The forces of degradation do not arise spontaneously. They are the result of interactions among the major agents in our societies. These agents are individuals, companies, and governments that relate through a variety of mechanisms and institutions such as families, firms, markets, governments, politics, clubs, universities, and online networks. Our focus here is primarily on impersonal relationships, such as those that occur in markets through buying and selling, through politics via laws and regulations, and through the social and antisocial activities of corporations.

Most of our actions, such as eating an apple, are neutral because they do not affect others. Others are beneficial, such as when we make charitable contributions to our schools or lend a hand in relief efforts. However, a wide variety of activities, such as pollution or overfishing, are harmful or Brown because they impose uncompensated costs on others.

These spillover effects are called *external economies* or *externalities*. They result from the impacts of economic activities that take place *outside of the market*. The most visible externality is pollution, such as the smog from automobile emissions that occurs in large cities or the dead fish, killed by toxic wastes, that wash ashore in lakes. Perhaps the least visible externality is the tiny coronavirus, a thousand times smaller than a grain of sand but even more dangerous than a runaway train.

The common theme of all externalities is that "the price is wrong," meaning that prices do not reflect social costs. This profound point can be understood as follows. For well-functioning markets, the recipients pay for the benefits of the goods they enjoy, and the producers receive the costs of the goods they produce. For activities with important externalities, the costs, benefits, and prices are not properly aligned. In the case of city smog, those who drive the automobiles are not paying for the health damage to those who breathe the harmful air. The Green movement devotes much time to analyzing the sources, mechanisms, and impacts of these spillover or externality activities.

FEDERALISM IN RESPONSIBILITIES

As in many areas, a central principle for dealing with externalities is federalism. This means that responsibilities should be lodged at appropriate levels of the societal hierarchy—personal, family, organizational, governmental, and global.

In other words, when considering the remedies for externalities, we need to ask which governance structure is best to deal with each one efficiently. Federalism recognizes that legal, ethical, economic, and political obligations and processes operate at different levels, and the solutions will necessarily involve various institutions and decision processes depending on the level. Moreover, the norms at one rung of the federalist ladder will differ depending upon how well the other rungs are performing. This is what is called *Green federalism.*

Take air pollution as an example (this discussion fits sulfur dioxide neatly, but other cases would be analogous). In an unregulated environment, companies would pay nothing when they pollute because the price of emissions is zero. However, each ton of emissions from a power company might cost residents of the nation $3,000 of health and property damage. Hence, there is a mismatch between the costs to polluters and the impacts on residents.

We can consider dealing with such air pollution with rules at different levels—individual, town, state, company, nation, or world. History shows that five of these six are ineffective, and only national regulation is effective. Individuals have weak incentives and poor information. At the other extreme, the United Nations has no mandate to control national pollution. So the most effective place to deal with air pollution is at the national level, and that indeed has been where most of the action has been found.

A particularly thorny issue involves activities that must rise to the top rung of the federalist ladder—global issues such as climate change, ocean pollution, and pandemics. For these issues, the global institutions and mechanisms for control are weak or nonexistent. It is not surprising, then, that many of the major failures to control the forces of degradation are on the global scale.

BASIC PRESCRIPTIONS FOR POLICIES

If the major challenges in creating a Green society are the threats arising from a range of externalities, the most effective policies are to "internalize" the costs and benefits. Internalization requires that those who generate externalities pay the social costs. Justice also requires compensation for those who are harmed.

The most harmful spillovers result from market transactions—importantly, those in the energy, transportation, and natural-resource sectors. Green policies will involve governmental actions to align private actions with public interest. These include regulations, taxes, liability laws, improved property rights, and international coordination of policy. Other inefficiencies result from behavioral anomalies, such as short time horizons, poor information, or laziness. Behavioral issues are more complex, but at the very least, they require improved information.

To the extent that countries fall short of the ideal well-managed society because governments fail to take the requisite actions, as all do to some extent, this imposes responsibilities on agents at different levels. The prescriptions for actions at various rungs of the federalist ladder will depend upon the extent of the uninternalized spillover and the effectiveness of different institutions.

For example, corporations and other private institutions may need to step in where governments fail. An important new development is corporate social responsibility, which directs corporations to act ethically in areas of their special expertise, such as the safety of their products and processes. For example, pharmaceutical companies need to inform doctors and patients about the health risks of their drugs—not deceive them and contribute to tens of thousands of overdose deaths, as Purdue Pharma did with promoting opioids. Individuals also have the responsibility to prevent unnecessary harm from their actions.

MECHANISMS

Societies will deploy different mechanisms to deal effectively with spillovers. These will be market incentives, governmental regulations and fiscal penalties, organizational activities through corporate responsibility, and personal ethics for important interpersonal interactions.

Examples of Green mechanisms include individual actions such as minimizing wasteful energy use, thereby limiting various pollutants. Additionally, they entail Green governmental laws and regulations, such as those that reduce emissions from power plants and automobiles. Another important Green activity is improved corporate management—running companies to take account of harmful impacts on their workers and customers. Additional actions include Green design—for example, the invention of new products that degrade quickly and innocuously at the end of their useful lives. *In short, Green actions are ones that tilt the social playing field away from harmful interactions and toward beneficial ones.*

Many of the most important spillovers require governmental actions. Actually, the first policies to deal with externalities are aimed at protecting the public's health against contagious diseases. We will see in chapter 11 that pandemics go back to the earliest written records. The word "quarantine" that is now familiar to people had its origins in 14th-century Venice, deriving from *quarantena*, meaning "forty days." In an effort to protect its citizens from the plague, Venice required ships to anchor for forty days before landing. Today, people must isolate for fourteen rather than forty days, and now it is sometimes on cruise ships.

More familiar are policies to reduce air pollution. In the first American century, air pollution was ignored. If it rose to the level of real damage, it was dealt with by litigation, generally as a private nuisance. This was seen to be ineffective, and state and local governments took the next steps, declaring air pollution a public nuisance and requiring factories to reduce smoke as early as 1881.

The major federal legislation on air pollution was not enacted until 1970. That law extended rules to all harmful air pollutants, but actions still leaned primarily on technology standards, which mandated specific technological fixes for pollution. The era of market instruments emerged with the development of tradable emissions permits for sulfur dioxide in 1990, and several countries enacted pollution taxes (such as carbon taxes) in the 2000s. Treaties and other agreements, which are like contracts between countries, cover the regulation of international air pollution. The history of air pollution, then, covers most of the major instruments that can be used to control externalities.

Air pollution and climate change are examples of extreme externalities, in which virtually all the damages flow to parties other than those generating the externality. Climate change is the pole of poles. If I emit a ton of CO_2 when I drive my car, 0.00001% of the climate damages flow to myself ("own"), while 99.99999% of the costs land elsewhere (on "others"). Damage to others is to other people, other lands, and other generations.

Some cases, such as common-resource problems, are more complicated because they have a more balanced mixture of own and other benefits. Take the example of congestion. Most people fume when they are stuck in a traffic jam. But they are likely to ignore their impact on other drivers. The result is that people may resist the obvious mechanism to reduce inefficient congestion: congestion pricing of highways and airports.

Another set of special mechanisms lies in the field of Green ethics and applies, as we shall see, not only to companies and individuals but to finance and even chemistry. Particularly important is corporate social responsibility, in which corporations need to be required to provide better information about the safety of their products and processes.

Closing Words

The metaphor of Green is inspired by the environmental movement that began more than a century ago and is among the most prominent and sustained efforts to contain harmful spillovers. However, the Green movement has spread far beyond the environment, and this is the story told in this book. Green thinking can help analyze and perhaps solve many of the thorniest problems of our age—global warming, pandemics, myopic decision-making, overpopulation of people, and overharvesting of forests and fish. It is also a good management tool for households, businesses, universities, and governments.

We now examine the role of Green in several areas and suggest how concepts developed by Green thinkers can improve the health and happiness of our increasingly interconnected world.

Green Efficiency

Efficiency is the staple diet of economists, who eat it for breakfast, lunch, and dinner.[1] But, as a society, our meals are sometimes spoiled by pollution, and this is also a major concern of economics.

The Great Lakes of North America are among the wonders of the natural world. They are the largest lakes and contain one-fifth of the world's freshwater. They were carved out of the land by retreating glaciers about 15,000 years ago. If you stand on the shore, you will see a vast expanse of blue water dotted with sails (or perhaps ice with ice fishing in the winter).

When humans began to industrialize, the Great Lakes became a giant dump for wastewater, factory pollution, and pesticides. Lake Erie, the smallest of the lakes, was declared "dead" because of algae growth, oxygen depletion, and massive fish kills. It suffered one particularly dramatic incident in 1969 when the Cuyahoga River, which runs through Cleveland into Lake Erie, caught fire. The uproar contributed to the passage of the Clean Water Act of 1972, as well as the U.S.-Canadian Great Lakes Water Quality Agreement in the same year.

These provide vivid examples of the problems confronted by the Green principles outlined in the last chapter. They demonstrate how poor economic management leads to the wasteful use of our natural resources. This point goes to the core of environmental economics, which is the subject of this chapter.

Recall that the spirit of Green has the goal of a well-managed society. This requires the efficient use and distribution of goods and services, both market and nonmarket, across people and time. By distribution, we mean the fairness of how goods are divided among populations.

This chapter focuses on Green efficiency, which draws on the concept of efficiency from mainstream economics. Indeed, we need invoke no new economic principles in pursuing Green efficiency. Rather, it focuses on a particular set of failures—those that primarily relate to dysfunctions in markets for environmental services and natural systems.

Promoting Green efficiency involves three central themes. The most important is the need to deal with the negative external effects of economic activity like pollution. A second concerns informational deficiencies, either consumer ignorance about energy use or insufficient innovation in technologies to promote Green behavior. The final issue involves dealing with *behavioral anomalies*, or inefficiencies that occur when people, firms, or governments act in ways that are not in their own best interests.

Background Philosophical Principles

Let us begin by sketching the underlying ethical principles commonly used in economics. A central principle is that social rankings are *individualistic*. In other words, social states are judged on the basis of how individual members of society rank them. If all individuals prefer world *A* over world *B*, then we will defer to their preferences.

This innocuous-sounding principle has an essential implication, which is the *Pareto rule*, named after an Italian economist of the early 20th century. This rule holds that if at least one individual likes social state *A* better than state *B*, and no one dislikes *A* relative to *B*, then state *A* is the preferred social state. The Pareto rule is a key reason why economists write so often about the role of markets in economic efficiency. In certain narrow circumstances, market outcomes cannot be bested by any other outcome according to the Pareto rule.

We begin with two background principles—individualistic rankings and the Pareto principle—that allow some progress in dealing with the efficient use of a society's resources.

Efficiency in General

What is efficiency? Efficiency denotes the most effective use of a society's resources in satisfying people's wants and needs. More precisely, economic efficiency requires an economy to produce the highest combination of the quantity and quality of goods and services given its technology and scarce resources. This is sometimes described in terms of the Pareto criterion mentioned above. In that language, an economy is producing efficiently when no individual's economic welfare can be improved unless someone else is made worse off.

Figure 4-1 will illustrate the discussion. Suppose that 1,000 perishable fish have been caught in a remote fishing village. One efficient outcome (call it equal *A*) would be that each of the 100 families gets 10 fish. But an alternative efficient outcome (call it unequal *B*) would be that one family receives 901 fish, and each of the others gets 1 fish. Unequal alternative *B* would hardly look fair to most people, but it is efficient.

An unfortunate *inefficient* alternative comes when disputes arise. Suppose that, because of the unequal alternative outcome in *B*, the citizens cannot agree about the distribution of the fish. They argue and debate about a fair procedure, and the negotiations drag on for days. They eventually decide to divide the fish equally. However, by that time, half the fish are rotten, so each gets only five fish. This is outcome *C*. True, most of the families are better off. However, the outcome is inefficient because of the wasted fish. This example also shows a potential trade-off between fairness and efficiency. If devising fair outcomes is costly, that may lead to inefficiencies. Figure 4-1 shows the three cases.

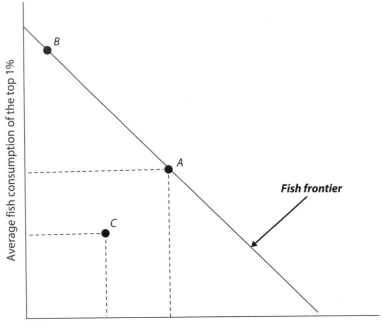

FIGURE 4-1. Two efficient outcomes and one inefficient outcome
The figure shows the average fish consumption of families in the bottom 99% and the top 1%. The outer line is the "fish frontier" of efficient allocations of the 1,000 fish. Point *A* is an efficient outcome with equality, whereas point *B* is efficient with high inequality. *C* is equal but inefficient.

All this discussion has been outside any institutional structure. Here is where economics enters. The central economic premise behind environmental economics is that markets allocate resources well when they function properly, but they can misallocate resources when there are market failures.

I will give this a colorful name by calling it "the invisible-hand principle," which refers to the efficiency of well-functioning competitive markets. This was the second pillar of the well-managed society discussed in earlier chapters. Adam Smith put the point eloquently in *The Wealth of Nations*:

> Every individual endeavors to employ his capital so that its produce may be of greatest value. He generally neither intends to promote the public interest, nor knows how much he is promoting

it. He intends only his own security, only his own gain. And he is in this led by an invisible hand to promote an end which was no part of his intention. By pursuing his own interest he frequently promotes that of society more effectually than when he really intends to promote it.[2]

What is the meaning of this enigmatic passage? Smith saw something that was not fully understood for almost two centuries after he penned those words—he saw that private interests can lead to public gain when they operate in well-functioning markets. More precisely, *modern economics has shown that under restrictive conditions, a perfectly competitive economy is efficient.*

Looking at figure 4-1, ideal competitive markets will ensure that society is on the fish frontier, such as with points *A* or *B*. These outcomes may be relatively equal or terribly unequal, but it is not possible to improve the economic status of everyone.

After two centuries of experience and thought, economists now recognize the limited scope of the invisible-hand principle. We recognize that inefficiencies might lead society to be inside the fish frontier, as in *C*. We know that there are "market failures," which occur when markets have defects. One set of market failures concerns imperfect competition or monopoly. Among the most famous monopolies in history were Standard Oil, American Tobacco, and AT&T. All of these were found to have engaged in unlawful practices and were dismantled. The giant technology companies today find the antitrust authorities nipping at their heels, complaining about their anticompetitive abuses.

Two other failures are keys to understanding Green efficiency. One concerns informational deficiencies, such as when people do not know the energy usage of different cars or appliances. However, most important failures come when spillovers or externalities occur outside the marketplace—negative spillovers such as pesticides discharged into Lake Erie or carbon dioxide (CO_2) emissions into the atmosphere.

The key point is that when any of these elements occur, Adam Smith's invisible-hand principle breaks down and surgery is needed.

Arthur Pigou: Founding Father of Environmental Economics

An earlier chapter recounted the founding of environmentalism by major figures such as Gifford Pinchot, John Muir, and Rachel Carson. The analytical thinking behind the Green movement originated with Arthur Pigou (1877–1959), a don at the University of Cambridge in the United Kingdom. He was educated at King's College, Cambridge, and became a leader of the Cambridge School in the early 20th century.

Pigou's life was devoted to developing economics as a tool for improving human welfare, and he had little interest in anything else—politics, foreigners, or women. It was said that Pigou was willing to accept corrections only from Cambridge or King's College economists and not from Americans. However, his biographer held that to be erroneous, stating that he was unwilling to accept corrections from anyone.

His major work, *The Economics of Welfare*, broke from the earlier tradition of debating conflicting systems (socialism vs. capitalism) to devising methods of improving the existing economic system.[3] While following in the footsteps of Adam Smith in believing that properly functioning markets would maximize human satisfaction, he saw clearly the flaws in the English economy of his time.

In his view, the major flaw was the presence of externalities, and he was the leading developer of that concept. Here is his description:[4]

> The source of the general divergences between the values of marginal social and marginal private net product that occur under simple competition is the fact that, in some occupations, a part of the product of a unit of resources consists of something, which, instead of coming to the person who invests the unit, comes instead as a positive or negative item, to other people.

His analysis uses the clumsy phrase "divergences between the values of marginal social and marginal private net product." This is exactly what is now called *externalities*.

Pigou provided several examples. One was the familiar case of lighthouses, which are "enjoyed by ships on which no toll could be conveniently levied." Other cases were parks, forests, roads and tramways, pollution control, alcoholism, and road damage. Some examples, which he took seriously but seem far-fetched today, included rabbits overrunning a neighbor's yard.

As we will emphasize in our chapter on innovation, the most important externality, in Pigou's view, was investment in new knowledge:

> [The most important externality comes from] resources devoted alike to the fundamental problems of scientific research, out of which, in unexpected ways, discoveries of high practical utility often grow, and also to the perfecting of inventions and improvements in industrial processes. These latter are often of such a nature that they can neither be patented nor kept secret, and, therefore, the whole of the extra reward, which they neither be patented nor kept secret, and, therefore, the whole of the extra reward, which they at first bring to their inventor, is very quickly transferred from him to the general public in the form of reduced prices.

While environmental economics tends to be a gloomy subject, emphasizing pollution and congestion, Pigou was right to point to the positive externalities of knowledge, invention, and improved technologies. The role of technologies in pursuit of the Green society will be discussed throughout this book and highlighted in the chapter on Green innovation.

Pigou's revolutionary proposal was to use fiscal tools to correct these externalities. Here are his words:

> The State [may], if it so chooses, remove the divergence [between marginal social and marginal private product] in any field by "extraordinary encouragements" or "extraordinary restraints" upon investments in that field. The most obvious forms which these encouragements and restraints may assume are, of course, those of bounties and taxes.

Pigou provides several examples of such fiscal measures. These include gasoline taxes devoted to developing roads, alcohol taxes to discourage excessive drinking, taxes on building in congested neighborhoods, and taxes on businesses in areas with high rates of disease.

Pigou's ideas gradually spread through economics and were advocated as a tool for using market approaches to reduce pollution. Pigou recognized the process but never used the term *externality*. The first exposition of the concept was by Francis Bator many years later, in 1957, after which it became widely used in economics and spread to environmental thinking and law.[5]

The environmental ideas of Pigou were at the time extraordinarily radical. Even before most economists acknowledged the damage that externalities worked on the economy, Pigou not only recognized them but also placed them into the standard economic framework and then devised a new remedy in the form of environmental taxes and subsidies.

It is useful to pause to celebrate the importance of this unusual kind of invention. Society celebrates inventions of new products like the Xerox machine or the smartphone. But many of the most important innovations are *institutional*. Our political democracy, invented in the 18th century, imperfect as it was, proved one of the most durable and valuable of all institutional innovations. Similarly, markets were invented and did not just spring up from the ground. And, as a last example, we can take environmental taxes and subsidies as a profound and important institutional idea.

We will return to the topic of pollution taxes in the chapter on Green taxation, but before moving on, we give a bow to the originator of this powerful idea.

Public versus Private Goods

A key economic distinction for understanding Green issues is that between public and private goods. *Public goods* are activities whose harms or benefits spread widely around the community, whether or not individuals pay for them or desire them. Private goods, by contrast, are those that can be divided up and provided separately to different individuals, with no external benefits or costs to others.[6]

A classic case of a public good is national defense. Nothing is more vital to a society than its security. However, national defense, once provided, affects everyone. It matters not at all whether you are hawk or dove, old or young—you will be living with the same military policy as the other people in your country. However, public goods differ completely from *private goods* like bread. Ten loaves of bread can be divided up in many ways among individuals, but what I eat cannot be eaten by others.

Note the contrast: The decision to provide a certain level of a public good like national defense will lead to spending and conflicts that will affect everyone without their individual decisions or consent. By contrast, the decision to consume a private good like bread is an individual act. You can eat four slices, or two, or none; the decision is purely your own and leaves others to eat what they want.

Lighthouses as Public Goods

Everyone but an anarchist agrees that national defense is a public good. A more illuminating example is lighthouses, which have an interesting history. Lighthouses save lives and cargoes. But lighthouse keepers cannot reach out to collect fees from ships nor, if they could, would it serve a social purpose to exact an economic penalty on ships that use their services. The light can be provided most efficiently free of charge, for it costs the same to warn 1 or 10 or 1,000 ships about dangerous rocks.

The two key attributes of a public good are that (1) the cost of extending the service to an additional person is zero (*nonrivalry*), and (2) it is impossible to exclude individuals from enjoying it (*nonexcludability*). Both characteristics apply to lighthouses.

But a "public" good is not necessarily publicly provided. It might be provided by no one. Moreover, just because it is privately provided does not indicate that it is efficiently provided or that the fees collected are sufficient to pay for the lighthouse.

The most spectacular musical of our era is *Hamilton*, which narrates in song the brilliant and tragic life and career of the nation's first public economist, Alexander Hamilton. A little-known side of Hamilton is that he sponsored the first public works, or infrastructure project, of the country, the Lighthouse Act. More completely, this was An Act for the Establishment and Support of Lighthouses, Beacons, Buoys, and Public Piers of 1789. Long before economic theorists developed the theory of externalities, Hamilton sang their praises as means of "the necessary support, maintenance and repairs of all lighthouses, beacons, buoys and public piers erected, placed, or sunk before the passing of this act, at the entrance of, or within any bay, inlet, harbor, or port of the United States, for rendering the navigation thereof easy and safe."

Lighthouses are no longer a central issue of public policy today and are mainly of interest to tourists and economists. They have been largely replaced by the satellite-based global positioning system (GPS), which is also a public good provided free by the government.

But lighthouses and national defense remind us of the key role of activities that cannot be efficiently provided by a free-market solution. These are not isolated examples. When you think about vaccines, pollution abatement, clean drinking water, highways, parks, space exploration, fire departments, or similar government projects, you generally find elements of public goods involved. The key point for Green policies is this: private markets are key to the efficient provision of private goods, but public goods require government interventions.

This fundamental economic principle will inform discussions of environmental policy from ancient problems like water rights to modern debates about the Green New Deal, which we will discuss later in this book.

Network Externalities

A completely different kind of externality arises in the presence of *networks*. Here is the basic idea. Many products have little use by themselves and generate value only when used with other products or other people. A network is such a product—one in which different people are linked together through a system. In an earlier era, important networks included physical linkages such as telecommunication systems, electricity transmission networks, pipelines, and roads. Increasingly today, the linkages are virtual, such as when people use smartphones, social media, and computer software or speak the same language (such as English).

To understand the nature of networks, consider how far you could drive your car without a network of gas stations or how valuable your cell phone would be if you were the only phone user. Similarly, credit cards and ATM cards are valuable because they can be used at many locations. Facebook caught people's fancy because they could network with others, and the more people on Facebook, the greater the benefit.

Networks are a special good because consumers derive benefits not simply from their own use of a good but also from the number of others who adopt the good. This is known as a *network externality*. Network externalities arise when the users of a good or service gain (positive) or lose (negative) when additional users adopt it.

When I get a phone, I can communicate with everyone else with a phone. Therefore, my joining this network leads to positive external effects for others. The network externality is the reason why many colleges provide universal e-mail for all their students and faculty—the value of e-mail service is much higher when everyone participates.

Networks also sometimes cause negative externalities. You may have been caught in a terrible traffic jam when the road got congested with too many vehicles for the size of the highway. Or perhaps you have sat on a crowded runway because the air network was full. Sometimes, computer networks can become burdened as well, slowing down people's service. These are all the opposite of the Facebook and telephone examples since a road is a network in which more people make the service less rather than more attractive.

Economists have discovered many important features of network markets. First, network markets are *tippy*, meaning that the equilibrium tips toward one or only a few products. Because consumers like products that are compatible with other people's systems, the equilibrium tends to gravitate to a single product that wins out over its rivals. One key example is Microsoft Windows, which became the dominant system in part because consumers wanted to make sure their computers could operate all the available software. Today, Windows has a market share of more than 80% of desktop operating systems.

A second interesting feature is that "history matters" in network markets. A famous example is the QWERTY keyboard used with your computer. You might wonder why this particular configuration of keys, with its awkward placement of the letters, became the standard. The QWERTY keyboard in the 19th century was developed in the era of manual typewriters with physical keys. The keyboard was designed to keep frequently used keys (like E and O) physically separated in order to prevent them from jamming.

By the time the technology for electronic typing evolved, millions of people had already learned to type on millions of QWERTY typewriters. Replacing the QWERTY keyboard with a more efficient design would have been both expensive and difficult to coordinate. Thus, the placement of the letters stays with the obsolete QWERTY on today's English-language keyboards, such as the one that I am using for this book.

The QWERTY example shows how a technology with strong network effects can be extremely stable. This point has important environmental consequences. America's automobile culture—with its existing network of cars, roads, gasoline stations, and residential locations—is deeply embedded and will be difficult to dislodge in favor of more environmentally friendly alternatives, like denser cities and improved mass transit. Other countries with different histories are less dependent on cars and roads. Indeed, one of the major challenges in making a transition to a Green energy structure is overcoming the existing network of energy-using capital and infrastructure. For example, perhaps electric cars will be the Green alternative. In that case, "gas stations" will need to be replaced by fast-charging "electricity stations" and fast-charging electric vehicles.

Network externalities raise different kinds of policy questions than older externalities such as pollution or contagious diseases like

COVID-19. Congestion is an increasing feature of our daily lives, particularly for those who live in cities or travel frequently.

One solution is to "internalize" the externality by having private ownership and operation of the network. If a company owns the entire network, it is motivated to build and operate a system that minimizes network spillovers. Cell-phone networks can suffer from congestion in busy times. But highly profitable cellphone providers have powerful financial incentives to invest in more capacity or use fancy pricing to reduce congestion at the busiest times.

It is more challenging to manage public networks, like roads and airways, because they operate by political consensus rather than profit maximization. We will return to the idea of *congestion pricing* and *congestion taxes* as innovations to make the most effective use of public networks.

Pecuniary Externalities

Environmental concerns arise primarily from technological externalities. These are spillovers, like pollution, where the interaction occurs outside the marketplace. Another variant, important but little discussed in environmental theory, is *pecuniary externalities*— those that take place indirectly *through the marketplace.* These occur because economic actions affect the prices and incomes of other people.

There are few detailed studies of pecuniary externalities. Perhaps the most important examples are those in which factories close and production moves to regions with lower wages. This often happens in the United States as companies move from high-cost to low-cost states. More controversial still are cases of "offshoring" production to other countries. Yet other examples—less controversial but quantitatively larger—arise because of the *creative destruction* of innovation and market churn that occurs when new products like computers replace old products like typewriters.

Take the example of the closure of a textile factory. Job losses of this kind typically result in extended periods of unemployment, and the workers end up with lower pay at other jobs. An important study by economists Steven Davis and Till von Wachter presents estimates of the impacts.[7] They found that, after plant shutdowns, workers typically lose about 15% of their earnings over the next decade. Suppose the jobs pay $50,000 a year. Displaced workers would lose about $75,000 over the next decade. This is an example of a large pecuniary externality imposed on workers who lose high-paying manufacturing jobs.

Pecuniary externalities such as job losses are complicated from an aggregate point of view because some workers gain jobs at the expense of job losers. Indeed, the total economic impacts of creative destruction from innovations and international trade are generally positive for individual countries and for the global economy. The case of Walmart has been carefully studied, and evidence shows that the rise of large superstores has significantly increased the real income of Americans because they have reduced consumer prices.

But to the laid-off workers, the loss is as real as a fall off a cliff. When a factory shuts down and production moves to Mexico or Vietnam, the gains to consumers and to some anonymous faraway workers are little consolation. The antiglobalization movement of Donald Trump and others reflects in part the pecuniary externalities that lead people to think that foreigners are "stealing" their jobs.

We should not underestimate the importance of pecuniary externalities or market gains and losses from changing economic structures. But it is critical to recognize that these have different structures than technological externalities because they occur *inside the market*. Economists generally believe that the remedy for pecuniary externalities like job losses from international trade is not to build high tariff walls but rather to have adequate unemployment insurance and retraining programs. Similarly, halting the creative destruction of innovation in telecommunications, new retail strategies such as big-box retail stores, and Internet commerce would in the long run reduce the living standards of virtually everyone.

5

Regulating Externalities

Modern environmentalism is an intellectual and legal framework that analyzes the major externalities of human activities. Its scope covers the entire world, from the smallest villages to the largest counties. All major universities have programs in environmental science and policy, and students find it a rewarding and inspirational field of study. As a field, it encompasses earth science, biology, ecology, public health, economics, political science, law, and many other fundamental disciplines.

Externalities represent a *market failure*, which occurs when markets malfunction. For externalities, the failure occurs because those who produce the spillover (say, the pollution) do not pay the damages (done to humans and fish). In the cases of significant externalities, efficient functioning requires that governments step in to correct the malfunction by regulatory or fiscal means.

Governments are a vital part of the enterprise. At one end, governments fund much of the science, including remote sensing from satellites, public health, and computerized modeling. Governments have also enacted a network of laws and regulations. Virtually every aspect of the economy—particularly those involving interactions with air, water, land, or energy use—involves some kind of government regulation.

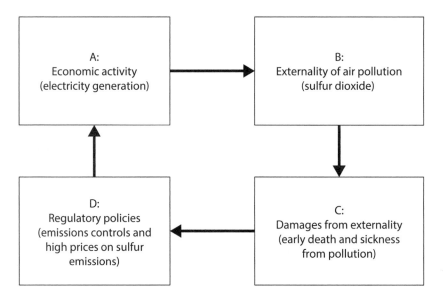

FIGURE 5-1. The circular flow of externalities and regulatory responses to air pollution from electricity generation

Regulating Externalities in a Picture

We can illustrate the logical structure of the production and regulation of externalities in a simplified way in figure 5-1. The problem begins in box *A*, with electricity generation, perhaps from coal. Combustion of coal has an unintended side effect of emitting a pollutant, sulfur dioxide, into the atmosphere, as shown in box *B*. The next step, shown in box *C*, is the impacts, which are the damages to human health from air pollution.

That would be the end of the story if there were no regulatory response. However, for most major externalities today, governments have a regulatory response, shown in box *D*. This involves steps to reduce emissions or internalize the externality. Governments can respond in several ways. Some are simple commands to reduce pollution or use particular abatement technologies. Others will tax or put a price on pollution. Whatever the mechanism, policies will tend to correct the externality.

Pollution regulation thus closes the circle by affecting the incentives of those who generate electricity. If the price of sulfur pollution is high, utilities might use low-sulfur coal, or they might add equipment to remove the sulfur, or they might shut down the coal-fired plant and build a gas-fired generator. Consumers feel the effects as well if the price of electricity rises, reducing electricity demand, further reducing emissions.

Figure 5-1 looks simple. But each of the boxes represents a complex and imperfectly understood system. For example, the sulfur emissions might be in the Midwest, but winds then take the pollution aloft, where it is transformed into other compounds and fouls the air in the East. It requires complex meteorology to determine who in the East is exposed to sulfur from the Midwest. An additional uncertainty is the health response of people to different concentrations of pollution. Statistical studies provide evidence, but the data are not based on controlled experiments, so the exact concentration-health relationship is unclear. Additionally, economists do not fully understand the costs of regulations, which are a significant factor in the cost-benefit calculus.

Figure 5-1 shows how Green government policies respond to important external effects—a point that is key to understanding the issues raised in the Green movements.

Externalities as Ownership Problems

Many externalities arise because of muddled ownership claims on public or "common-property" resources.

Take the case of the earth's atmosphere. While a country might claim its air space, there is no ownership of outer space or of the chemicals in the atmosphere that circulate without any legal restrictions. The atmosphere is the common property of living things. In most countries, the price of emitting substances into the atmosphere is zero, and the result is rising concentrations of greenhouse gases, pollution such as sulfur dioxide, and satellite debris circulating the globe.

A more subtle externality is overfishing in the oceans. This is another example of a misuse of a common-property resource. You might wonder where the external effect lies here since the fishermen own the boats, hire the labor, mend their nets, and run the risks of stormy weather. A closer look reveals that fishermen pay for catching the fish, but they do not pay for the effect of depleting the breeding stock. When a fisherman catches a bluefin tuna, this reduces the number of bluefin tuna that can resupply the tuna stock. If the stock is depleted sufficiently, the species will go extinct because they cannot find each other or do not produce enough fry to grow up to mate and keep the species viable. Overfishing is an externality because the value of the breeding stock is excluded from the fisherman's cost-benefit calculus and is therefore underpriced.

Solutions

Legal scholars tell us that the problem raised in each example above is imperfect property rights. That is, the climate, clean air, and breeding stocks are common-property resources that are mismanaged. For a common-property resource, everyone's business is nobody's business. Private decisions ignore some of the valuable aspects of a system, and decisions are therefore tilted to overproduce bads and underproduce goods.

Is there a "free-market solution" to externalities? In some cases, the externality might be corrected by changing property rights. Suppose you own a large pond full of trout that you allow people to catch for a fee. Since you own the pond, you also own the breeding stock. You have proper incentives to manage the pond so the stock is not depleted. You might charge a sufficiently high price to reflect the value of each fish in producing the next generation of fish so that you have a viable business in the coming years. The conversion of common property to private property has been an important factor in improving land management in many countries.

In some cases, society might decide that an asset or resource has public-good characteristics that make it inappropriate for privatization. For example, certain unique and irreplaceable assets like Yellowstone National Park should not be sold off to the highest bidder for use as an amusement park or mining site. Rather, it should be preserved and managed as a public resource for the enjoyment of current and future generations and as a unique natural wonder.

In other cases it is hard to see how creating private property can solve the problem. No one owns the climate, or clean air, or the ocean's fish. So there is no private owner to do the calculations that include the value of the climate, of the clean air, or of the bluefin tuna stock. Given both law and science—ocean fish are a fugitive asset—there seems little prospect that the legal status of these socially valuable public assets will change in the near future.

What is the remedy for misused common property resources? In cases where it is not possible to create property rights for mismanaged common-property resources, governments may need to step in with regulations or taxes. The government might limit CO_2 (carbon dioxide) emissions to slow climate change. And it might limit fishing through transferable fishing quotas. Governments need to inspect factories to make sure they are not dumping toxic wastes in streams and lakes. The list is long but not endless. The point is that markets work wonders when they have the right price incentives, but when there are important externalities and the prices are wrong, unregulated private markets can lay waste to the land and air and oceans.

Positive Externalities and Improved Technologies

The spirit of Green often shows a pessimistic face, fretting about the ills of pollution, climate change, and irresponsible corporations. However, we need to step back and recognize the powerful role of *positive externalities*. These primarily involve new knowledge and technological change but also include the development of institutions that can improve the human condition.

A useful antidote to gloomy outlooks is a review of the progress in health, longevity, and living standards. If we look at virtually all the indexes of living conditions, they have been improving steadily for the last two centuries. Global per capita income has grown at almost 2% per year since 1900. This economic progress has been accompanied by steady increases in life expectancy and declines in many dread diseases. Steven Pinker's astounding book *The Better Angels of Our Nature: Why Violence Has Declined* describes many social improvements over the ages.

The root cause of these improvements has been new scientific and technological knowledge. Virtually any product you can think of—vaccinations, smartphones, TVs, cars, copy machines, winter strawberries, and the Internet, to name a few—are the result of technologies developed over the years.

Moreover, economic studies of key technologies show that they also have substantial externalities—but positive spillovers in these cases. In each of the products listed in the last paragraph, the key inventors received at most a tiny fraction of the social benefits of their inventions. For example, Chester Carlson invented xerography and saved billions of hours of drudgery for scribes and secretaries, but he earned only one-sixteenth of a cent for each Xerox copy during the patent and earned nothing in the last half century. His is the typical case of an inventor who could not appropriate the benefits of his invention. Even tycoons like Apple's Steve Jobs, who definitely did not die penniless, reaped only a small fraction of the value of the iRevolution.

Economists have devoted much time to the study of promoting rapid and fruitful technological change. Indeed, the Nobel Prize in Economics of 2018 was awarded to Paul Romer for his pathbreaking work on the public good of knowledge. New processes and products are often the keys to attaining Green goals. To promote Green design requires attention to mechanisms such as properly pricing pollution, strengthening intellectual property rights, and providing government support for fundamental Green science. Many scientists think that the economic, social, political, and health crises caused by COVID-19 will only be solved when the population is fully vaccinated and people feel safe to return to their normal daily life. The major point here is to remember that the positive externalities from improved technologies and institutions can offset the negative externalities if nations make wise investments and choices.

The Fundamental Condition for Optimal Pollution

Once we have diagnosed the syndrome of externalities like pollution, it is a natural step to consider how to achieve "optimal pollution." This term may seem strange—how can any pollution other than zero be optimal? However, the term reflects the reality that reducing pollution is costly while the benefits are finite, so a balance of costs and benefits is usually necessary. It would be extravagantly costly to remove the last gram of particles from automobile emissions while removing the last gram will have a negligible impact on public health.

Therefore, the theory of optimal pollution aims at determining how many grams is the right balance between too much and too little. The economic and human stakes are high. Environmental regulation costs tens of billions of dollars each year, but it also produces substantial benefits (more on this later in this chapter).

The economic framework for determining the level of regulatory stringency is cost-benefit analysis. This means that regulations are set so that the costs of regulations are balanced by their benefits. More precisely, in an optimal framework, regulations are set so that the incremental cost of increasing the stringency of a regulation (or *marginal cost* in the language of economics) is just offset by the incremental benefit or damage averted (*marginal benefit*).[1]

Thus, suppose that regulators are studying the optimal standard for automotive tailpipe emissions for carbon monoxide (CO). They determine the costs and benefits of limiting emissions at different levels: 2 grams per mile, 3 grams per mile, and so forth. They determine that 3.4 grams per mile is the optimal standard. At that standard, scientists might estimate that the last ton of CO emissions causes $100 of damage. If this is the optimal or efficient regulation, then this last ton would also cost $100 to reduce.

The fundamental condition for efficient environmental regulation is that the marginal cost of reducing emissions equals the marginal benefit.

Table 5-1 provides a hypothetical example of optimal regulation. Consider what happens as the allowable level of pollution moves from 900 to 800 to 700 and so on. At a pollution level of 300, the cost of additional reductions just balances the extra benefits. Total costs plus damages are minimized. The socially optimal pollution is determined this way.[2]

TABLE 5-1. Optimal pollution

Pollution (tons)	Abatement (tons)	Total cost of abatement	Marginal cost of abatement	Benefits from abatement	Marginal benefit of reduction	Total benefits minus total costs
900	0	0	0	0		0
800	100	8	0.17	483	3.42	475
700	200	33	0.33	685	1.79	652
600	300	75	0.50	840	1.43	765
500	400	133	0.67	971	1.23	838
400	500	209	0.83	1,087	1.10	878
301	599	299	1.00	1,190	1.05	890.9819
300	600	300	1.00	1,191	1.00	890.9823
299	601	301	1.00	1,192	0.96	891.1000
200	700	409	1.17	1,287	0.93	879
100	800	534	1.33	1,377	0.87	843
0	900	676	1.50	1,461	0.82	786

At abatement of 600, the marginal cost of additional reductions just equals the marginal benefit. The last column shows that net benefits (total benefits minus total costs) are maximized at that level of abatement.

The fundamental condition allows us to circle back to Adam Smith's invisible-hand principle and to Arthur Pigou's theories of "extraordinary discouragements" as a way of understanding optimal environmental policy. In the ideal case of efficient markets, the marginal cost of producing bread (incurred by farmers and millers and bakers) just equals the marginal benefit of consuming bread (paid for by consumers). There is no gap between the marginal cost to society and the marginal benefits to society because there are no external effects.

Similarly, for automobiles, if all costs of automotive travel were *internal* to the car's owner, then (leaving aside other possible distortions) there would be no need to impose environmental regulations on automobiles. However, when externalities exist, the social costs of production (including the pollution) are larger than the social benefits of consumption. The difference between the marginal social costs and social benefits is exactly the impact of the externality. With no regulation, the first 100 units of abatement have net benefits of

475 in table 5-1. As pollution is reduced, the net benefits decline until, at the optimal level of 300 units, the marginal cost including pollution (almost exactly) equals the marginal benefit of the product.

The fundamental condition for efficiency applies broadly across many fields. It is illustrated here for pollution. Another highly consequential application is for climate change. To foreshadow the discussion of global Green below, modelers have estimated the marginal damage from emissions. This is approximately forty dollars per ton of CO_2, according to the most comprehensive estimate of the U.S. government. By setting the price of CO_2 emissions at forty dollars per ton, countries can ensure the appropriate balance of global abatement costs and global damage control.

We will return to this fundamental condition on several occasions in the chapters that follow. I emphasize that this condition is idealized and, like the perfect game in baseball, seldom seen in reality. Governments do not always use the fundamental condition to set policy, but in well-managed societies it is often consulted in the background to ensure that actual policies are close to efficient levels.

6

Green Federalism

Many political systems have a federal structure in the division of powers among national and subnational governments. A federal structure requires that the central government and the subordinate bodies have well-defined responsibilities and rights to manage their spheres of influence. For example, national governments are usually responsible for tariffs and national defense, while local governments are responsible for educating children and collecting trash. The division of labor is helpful because the political boundaries are often those of the public good, and political authorities at the different levels usually possess the specialized knowledge and political incentives to understand and solve collective problems.

Green Federalism

Policies for externalities also have a federal structure. *Green federalism* recognizes that legal, ethical, economic, and political obligations and processes operate at different levels, and the solutions will necessarily involve various institutions and decision processes depending on the level. Figure 6-1 illustrates different externalities and the relevant place on the federalist ladder where they are most effectively regulated. Climate change requires global coordination,

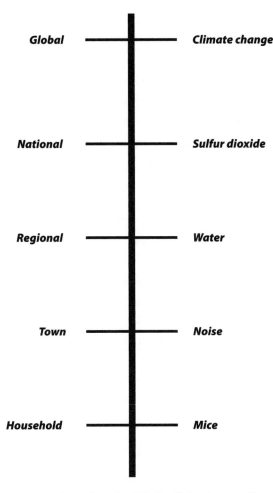

Global	— Climate change
National	— Sulfur dioxide
Regional	— Water
Town	— Noise
Household	— Mice

FIGURE 6-1. Externalities should be handled at the most effective place on the federalist ladder

while noise regulations are best handled in cities and towns, and households set mouse traps.

Environmental Federalism in the United States

Consider the questions of environmental policy for the United States. Most air and water pollution policies are covered by federal laws and regulations, such as those under the Clean Air Act of 1970 and many amendments. Provisions such as the limits on tailpipe emissions of automobiles are determined by the federal U.S. Environmental Protection Agency (EPA). States and tribal entities develop plans with EPA approval, and the lower entities monitor compliance.

Even though much regulation is at the federal level in the United States, states will often impose further restrictions on top of federal ones. For example, California has among the most stringent environmental protections. A 2015 law requires California to get half of its electrical power from renewable sources by 2030. Some states are less enthusiastic. For example, Mississippi is at the forefront of litigation *against* federal environmental standards.

Cities and towns are primarily involved in land use, including garbage disposal. Building codes are important to prevent fires and floods but also to ensure that minimum standards of housing are provided. Cities often limit nuisance activities. For example, the city of New Haven limits to six the number of chickens that a homeowner may own, and roosters are not allowed.

Other countries will draw the borders of Green federalism at different places. But the general principle of division of labor, with regulation placed at the point where it can be most effectively managed, is the thread that runs through these decisions.

Externalities Arising from Principle-Agent Conflicts

Many decisions in an economy depend upon teams working together—either cooperatively or selfishly. When we seek medical treatment, the team consists of doctors, the care team, the insurance company, and the government along with the patient and family. The doctor is an expert who suggests the treatments, but others pay the bills, receive the treatment, or comfort the sick.

Teams can interact cooperatively, as when a baseball team wins the World Series. Or they can work at cross-purposes and destructively. When the interaction is harmful, it is called the *principal-agent problem.* A more intuitive name is "the landlord-tenant problem."

Principal-agent relationships are ubiquitous in a complex, interdependent society and are a useful way to understand externalities and Green issues. Because they operate at different levels, they can also illustrate the issues of Green federalism.

Putting normal exchange and externalities into the principal-agent framework is helpful. A standard market transaction has a close linkage between those who enjoy a good (the consumers) and those who provide it (the firms) because consumers pay producers an agreed-upon price. When you buy a pair of shoes, you pay for the costs of producing them, and the manufacturers and retailers are compensated for their efforts. Thus, incentives are aligned. If both sides are well informed and there are no spillovers, principal-agent problems are unlikely to arise for market transactions.

Externalities by contrast are a pernicious principal-agent problem because the principals and agents are completely distinct, have different incentives, and often do not know each other. When producing the shoes requires the combustion of fossil fuel—to heat the factory that stamped the shoes or fuel the truck that delivered the shoes—one important cost is not covered: the damage caused by the carbon dioxide (CO_2) emitted. There is no linkage between the agent-polluter and the principal-pollutee.

What is the root cause of the principal-agent or landlord-tenant problem? It arises when the knowledge or incentives of those making the decisions (the agents or tenants) differ from those who experience the consequences of the decisions (principals or landlords). The distortions caused by principal-agent problems become increasingly severe when principals and agents are different people and have diverging values and incentives.

Landlords and tenants often have disputes because the landlord is interested in the long-term value of the property, while the tenant just wants a comfortable place to live for a year or so. Tenants are unlikely to engage in long-term upkeep or ensure a marketable property. A similar example is how people treat a rental car—badly. This point is expressed in the adage that no one in the history of the world has ever washed a rental car.

This syndrome also arises for publicly owned corporations. Here, managers are agents with incentives to pay themselves generous salaries, but those dollars come at the expense of dividends to shareholders who are principals. The principal-agent syndrome is undoubtedly one of the reasons for the skyrocketing executive compensation of recent years.

Federalism and the Principal-Agent Problem

The principal-agent problem is useful because it shows how externalities operate at different levels of society. Some involve households, while others operate at the local level, through institutions (such as corporations or universities) and perhaps through contractual relationships such as leases; some are national; and some are transnational or global.

Some principal-agent interactions involve *personal decisions*. As a student, how should I spend my time? Should I study or go to a party? I have aligned incentives because I am both agent and principal. If I as agent go to a party, I am the principal who has a low grade point average (GPA) and poor job recommendations. I have the incentives as an agent to do what I as principal want.

Or perhaps not. Sometimes, we make poor decisions. Perhaps we stayed too long at the party and overslept the exam. Or engaged in substance abuse and was in a daze for the exam. We might say that the party person was the agent whose incentives ruined the prospects of the student person. Later, in our discussion of behavioral perspectives, we will say that the present person underweights the importance of the future person. So here is a kind of personal principal-agent problem.

Another rung up the federalist ladder are *household decisions*, say, of a family. While families generally have shared objectives, conflicts sometimes arise. Often, the family principal-agent problem involves the fact that one person (agent) makes the decisions, while the other (principal) pays the bills. A common occurrence is that the agent-child forgets to turn off the lights while the principal-parent pays the bills.

Each family has its own approach to solving these problems as they arise. One solution to the lights problem is to give agents (children) a point when the lights are turned off and allow them to spend their points on a special treat. However, distortions are likely to be small at the family level because agents and principals have so many shared interests.

For most externalities, such as those at higher levels, there are no shared interests at all. This implies that the incentives are misaligned. One pervasive principal-agent distortion arises because of imperfect contractual arrangements, which is another example of the landlord-tenant conflict. Many apartment rentals require the landlord to pay for utilities, while the renter determines the energy use (temperature level, number of appliances, and so on). In those cases the agents are the people who rent apartments, while the principals are the owners who pay the utilities. Empirical studies show that the separation of decisions from fiscal responsibilities raises energy use substantially. A special case of this is student dormitories, where students face a zero price for energy use, while the college faces a market price, and the result is excessive use (relative to a well-functioning market).

Another class of decisions involves *governmental questions* at the local, state, or national level. Zoning questions, which involve congestion, noise, light, and green space, are important determinants of our local environment. Housing codes set the lower-bound requirements for housing design. At the next level are national public goods, such as clean air, national defense, pollution control, and other health and safety measures. Even more complicated principal-agent problems arise at the national level because the agents who make decisions (the legislators) are far removed from the principals (the people breathing the dirty air) and may care more about their party than the health of their constituents.

In no area is the principal-agent problem more evident than in the need to restrain the temptation of national leaders from going to war. George W. Downs and David M. Rocke express the principal-agent problem here clearly:[1]

> [The principal-agent] problem is particularly difficult in the areas of intervention and interstate conflict in which the chief executive . . . may possess preferences for or against participation in war that are different from those of the median constituency

member. In a democracy, the mechanisms that help deal with principal-agent problems range from a free press and legislative declaration of war to electoral defeat and impeachment. In an autocracy, there are far fewer of these mechanisms, and at the extreme, there may be nothing more than the costly option of armed rebellion.

We can think of the principal-agent conflict over war as one in which the agents (the leaders) command large armies, move pieces around on a map, and may be feted as glorious victors, while the principals (soldiers) get bogged down and shot at in the jungle or the desert.

It is well established that the greater the distance between the decider and the decidee, the greater the chance for misaligned incentives and unrepresentative decisions. However, as Downs and Rocke note, there are mechanisms at work, particularly in democracies, to place checks on unrepresentative decisions. Moreover, the history of environmental legislation in democracies suggests that the interests of the public get a voice over the longer run (as described in the later chapters on Green politics).

The top rung of the federalist hierarchy involves global externalities, exemplified by issues such as pandemics, global warming, and ozone depletion. In the principal-agent framework for climate change, the agents are the people driving and heating their homes today and emitting CO_2 in one country, while the principals, harmed by climate change, are different people, in distant times and lands, largely unborn, and unknown to agents.

We will see that the links between principals and agents for global externalities are so weak, and the mechanisms for repairing the adverse incentives so feeble, that problems here are the most difficult to solve.

7

Green Fairness

Two famous writers were discussing fairness. Scotty Fitzgerald is said to have said, "The rich are different from the rest of us," to which Ernest Hemingway replied, "Yes, I know, they have more money."

As an avid fisherman, Hemingway might just as well have said, "Yes, they have more fish." An earlier chapter discussed disparities in fish fairness among households. In that example a fishing village was considering how to allot 1,000 fish that had just been caught. Most people would agree that the fish should be distributed efficiently— that is, so they reach the homes of the population before rotting. But another pillar of the well-managed society is fairness. The fish and other goods and services should be apportioned fairly among the population.

This principle applies to environmental goods and services as well as fish and other goods. A fair society would ensure that everyone up and down the income ladder would enjoy clean water, healthy air, green space, public parks, and similar aspects of an environmental living standard.

What do we mean by fairness? Political and moral philosophers are deeply divided on this question, and political parties are similarly polarized. Economists generally refer to "inequality" rather than fairness because the former term can be measured in terms of

TABLE 7-1. Income distribution in the United States, 1967–2018

Year	Lower 20%	Middle 20%	Top 5%
		Level (2018$)	
1967	10,545	46,653	185,294
1990	13,390	55,649	259,281
2018	13,775	63,572	416,520
		Growth rate (%/yr)	
1967–1990	1.0%	0.8%	1.5%
1990–2018	0.1%	0.5%	1.7%

Source: Data are from the U.S. Census Bureau, https://www.census.gov/data/tables/time-series/demo/income-poverty/historical-income-households.html.

differences in income and wealth, whereas fairness is a subjective term that is not easily quantified.

The discussion of Green fairness that follows begins with general reflections on the sources and measures of inequality. It then focuses on questions that are emphasized in the literature on environmental justice, generational fairness, and fairness toward animals.

Measures of Inequality

Let us begin with an overview of economic inequality. There are extensive statistics on inequality in the United States and other countries. Table 7-1 shows the levels and trends of income for the United States over the last half century. It displays the average incomes of the bottom 20% of the population (the poor), the middle 20% (the middle class), and the top 5% (the rich). Two facts are clear. First, the rich have vastly more income than the poor. The average income of the top 5% was 30 times that of the poor in 2018.

A second feature is shown in the last two lines. The incomes of the top 5% have risen much more rapidly than the lower two groups. While the patterns vary slightly across the different time periods, incomes of the rich rose by 120% over the 1967–2018 period, while those of the middle and lowest group rose by only 30%. In the second interval (1990–2018), the poor had virtually no growth in income.

Sources of Inequality

Let us look behind the numbers in table 7-1 at the sources of inequality. Views on the fairness of the existing level of inequality might differ depending on whether high incomes were the result of effort or luck or inheritances. Just to simplify, we suppose that good luck and lots of effort lead to productive, long, healthy, and happy lives. Bad luck, bad neighborhoods, and poor efforts might lead to the opposite.[1]

Effort is shorthand for people acting purposively to meet their personal goals; it is clearly one critical ingredient in a good life. For some people, effort means years of training and working all hours of the night to achieve financial success or win a medal in the Olympics. Others may want to spend time with family or studying religious texts. Some may want to be ski bums or enjoy life on the road. Effort is well spent if it achieves life's goals.

Setting effort aside for the moment, we must emphasize luck because life is a giant lottery. Your life success depends on the randomness of your DNA, your country, your family, your teachers, and the state of the job market when you look for work.

One glaring violation of fairness is the lottery of location. People born in rich regions, who go to good schools and live in safe neighborhoods, generally find that Fortune shines on their incomes and health. A second major source of inequality is race and skin color. Here is how labor economist James Heckman describes the lotteries of distribution:[2]

> The accident of birth is the greatest source of inequality in America today. Children born into disadvantage are, by the time they start kindergarten, already at risk of dropping out of school, teen pregnancy, crime, and a lifetime of low-wage work.

Because the various characteristics of the lottery are determined by location, scholars have found a "zip code effect." If you were born in zip code 10104 in midtown Manhattan, your average annual income is $2.9 million. If, however, you live in a district of the South Bronx, just a few miles away, your average annual income is $9,000.

Inequalities in America are multiplied in other countries. If you happen to live in the middle of a civil war or in a country without a functional medical system, your life will probably be miserable and possibly be short. If we define fairness as equal lotteries, then life is definitely unfair. Moreover, there are literally walls that prevent those in war zones or the poorest regions from enjoying the safety and high living standards of rich countries.

In an earlier era, people often thought that their fortunes lay in the stars. Today, social scientists hold that fairness depends on whether public and private institutions and policies offset or reinforce the lotteries of life. Historically, social and economic policies often worsened the outcomes of lotteries. Those who started disadvantaged—women, indigenous peoples, and people of color in most of America—were subject to discriminatory practices, exclusion, expulsion, and slavery.

Today, in liberal democracies, most policies emphasize equality of opportunity, which essentially means that policies are neutral with respect to life's lotteries. You have one vote, whether you are rich or poor. Your tax rate depends on your income, not on your accent or skin color. Some policies, such as unemployment insurance, tend to offset the effects of bad lottery outcomes, and these programs are available to all who experience bad outcomes.

But progress is uneven. Black lives continue to be at risk long after slavery was abolished. We see that the gains in one generation are offset in the next when a pandemic cuts its swath through the population, killing hundreds of thousands and throwing tens of millions out of work while inept and uncaring leaders focus on their own economic and political fortunes rather than on the public interest.

Fairness in Markets?

A perennial question is whether the distribution of outcomes in a market economy has a claim to fairness. Look back to table 7-1. Scholars generally believe that market forces have been a major component of the growth gap between rich and poor over the last half century. These forces include technologies that substitute for low-skilled workers, as well as the trends of globalization, migration, and deregulation.

Was that fair? When we see that the top 1% of households today own 40% of U.S. wealth, is that fair? Is it the result of effort or luck? Do Americans earn 10 times what Africans make because they work so much harder? If the lottery of life gives one person a million-dollar income while another is barely earning the minimum wage, is it a fair outcome?

Judging fairness of market forces is a matter of values rather than pure science. Here is an analogy that can illustrate the ethics of the market. Think of the distribution of food in the jungle. Lions can eat pretty much anything they can catch, while antelopes can enjoy a plentiful diet as long as they can flee the lions. Would we say that the law of the jungle is ethically just? Probably not. Human societies are more civilized than the jungle, but the food distribution under pure laissez-faire has a similar ethical status to the distribution of spoils in the jungle.

What Is Different about Green Fairness?

Does Green fairness add anything to the more general considerations? We can point to three further concerns. The first concerns generational fairness or how we treat the future. Second is environmental justice, including the impact of environmental policies on the distribution of income. Finally, environmental ethics adds an entirely new dimension in asking about fairness to animals. We take up these topics in the balance of this chapter.

Generational Fairness

A first concern relates to future generations. Is it fair to the future to leave a planet under siege, with rising oceans, threatened extinctions, greater weather extremes, and damaged ecosystems? But is it perplexing here to know who is injuring whom? The injustice is hazy because it is inflicted by billions of people on billions of other people, where none can be pointed to as the single guilty party.

Here is a way to think about generational fairness. If you could press the "restart" button on your life, when would you choose to be born? Would you prefer exactly the period you are living through? Would you like to live in the long-haired 1960s? Or perhaps in the future?

We might prefer the future because medicine is improving rapidly, and robots will be around to wash the dishes and tend to our every whim. Or we might prefer today's world as humanity's golden age if we fear that the future will be plagued by autocratic robots, cyberwarfare, and a deteriorated natural world. If people would prefer to be born in the future rather than in the present (born in 2050 rather than 1990), then it is hard to argue that we are unfair to future generations. Here again, however, generational fortunes are a combination of effort and luck, so if we prefer the present, it might be because of luck rather than generational discrimination.

The issues of generational fairness are profound ones that occupy much thinking in the Green movement. These will be taken up in greater depth in the next chapter's discussion of sustainability.

Environmental Impact on the Distribution of Income

A second issue is the impact of environmental degradation and corrective policies on the distribution of income, or, more generally, on the distribution of economic welfare. We first tackle the question of environmental justice and then look at the broader distributional issue.

ENVIRONMENTAL JUSTICE

Environmental justice is narrowly defined as equal access to the development of environmental laws and regulations for all people independent of race, color, national origin, or income. In the language of the lottery, all should have equal *opportunity* in the environmental lottery. For the spirit of Green, a broader definition is an equal distribution of environmental benefits and burdens, which suggests that lottery *outcomes* should be equalized.

Here is an example: Central Park in New York City is one of the greatest urban parks in the world, and it is the most heavily subsidized park in that city. Who benefits the most? The main beneficiaries are people living near the park, some of the richest people in the world. How fair is that? Should New York City allocate more of its spending to poorer zip codes in the Bronx?

A closer look contains a few surprises. People who live near the park are richer, but they also pay a huge premium for that benefit. One study found that you would need to pay $1½ million more for an apartment near Central Park than in more distant neighborhoods. Interestingly, this was one of the reasons advocated by the park's designer, Frederick Law Olmsted, who argued that the increased property taxes on bordering properties would more than pay for the park.

Other examples of injustice are more compelling. City planners often put parking garages and garbage dumps in low-income neighborhoods. The rationale is that land prices are lowest here. But this calculation is flawed. It leaves out both nonmonetary consequences (such as health) and monetary transfers (such as losses in property values to residents, who are probably least able to afford the losses). Increasing air pollution in a low-income neighborhood imposes a burden on a population that is already relatively unhealthy and has poor health care. The precept of including all costs—nonmarket as well as market—for projects that have the potential to cause harm is a sound principle for fairness as well as for efficiency.

The Distributional Impact of Environmental Policies

Are environmental policies themselves currently fair? More precisely, is the structure of regulatory costs and environmental benefits one that tilts toward or against poorer households? Are policies regressive or progressive? (Regressive means that the impact lowers the economic welfare of poor relative to rich, while progressive means the opposite.)

The answer here is complicated because it involves indirect measures of costs and benefits as well as imputations by income class. However, the balance of the evidence suggests that the costs of abating pollution are regressive, while the benefits of the environmental improvements are progressive.

An example of the regressive nature of environmental costs can be seen with the gasoline tax, for which we have extensive data. A study by Antonio Bento and coauthors collected data on incomes and gasoline usage. It examined the impact on lifetime incomes of the four income groups (low, low middle, upper middle, and high). Setting aside the uses of revenues (and returning to this momentarily), the net impact of the tax was clearly regressive. The largest negative impacts were on the lower two income groups.[3]

The conclusion about the regressive nature of environmental policies has been found in many other areas as well. That is, lower-income groups spend relatively more on goods that are subject to environmental regulation (gasoline, electricity, and heating) than do higher-income groups. They spend *absolutely less* than the rich but *relatively more*. Therefore, the real incomes of the lower-income groups are generally reduced more than those of upper-income groups.

However, the regressive impact is not inevitable, especially when the environmental policies are of a fiscal nature (say, gasoline taxes) rather than a regulatory nature (such as emissions limitations). If the gasoline tax were recycled in a progressive manner (with larger rebates to low-income households), it could be turned into a neutral or even progressive program. Similarly, if climate policies are implemented through carbon taxes rather than emission restrictions, the taxes will raise revenues that can be recycled to the most heavily affected groups.

Most studies of the distributional impacts of environmental policies focus on abatement costs, such as the costs of reducing air pollution. But a complete picture would require including the environmental benefits as well. Continuing the example of air pollution, the

U.S. Environmental Protection Agency (EPA) estimated that the nation spent about .5 % of gross domestic product (GDP) annually to reduce air pollution from automobiles, power plants, and other sources from 1970 to 1990.

The EPA also estimates that the aggregate benefits of clean-air regulations over the same period were 40 times the costs.[4] The evidence on the distribution of the *benefits* of environmental policy is scattered but persuasive. There is a strong negative association between exposure to pollution and per capita income. To begin with, the evidence indicates that air pollution, as well as the location of toxic release polluters, are disproportionately located in neighborhoods with low income and high proportions of black and Hispanic residents. For example, one study divided cities into a more-polluted and a less-polluted half. It found that race, ethnicity, and income were strongly correlated with being in the polluted half of the city.[5]

The association of pollution exposure and income implies that the benefits of environmental policies are progressive. Since the poorer neighborhoods are more polluted, reducing exposure will have a larger effect on poor households. This is reinforced because poor households are likely to have inadequate health care.

In sum, when considered in isolation, the costs of pollution-abatement programs tend to be regressive, with a higher burden at the low end of the income distribution. The use of emissions charges or taxes rather than quantitative regulations generates revenues that can be used to offset the regressive nature of environmental policies. However, the impacts of environmental improvement to health and welfare appear to be progressive, helping low-income households to a greater extent. The net impact is unclear, but since the benefits tend to far outweigh the costs, the best guess is that the overall impacts of environmental policies are progressive.

Fairness to Animals

A third area that is especially important for Green fairness is proper consideration of nonhuman species, or fairness to animals. Economics, law, and moral philosophy usually include only the preferences or welfare of humans. However, there are exceptions, and views on the rights and welfare of animals are evolving.

Do animals have any legal rights? For the most part, the answer is no. Animals may have "interests," but they do not have "rights." What is the difference between rights and interests? Animal rights would mean that animals, like humans, have activities and status that cannot be sacrificed or traded away just because they might benefit others. Interests, by contrast, are protections, but interests can be balanced or compromised in return for other interests, although the exact nature of the trade-off is the subject of a lively debate.

Animal interests protect them against cruelty, which is unnecessary, and in some cases provide special protection, such as to endangered species. Moreover, most laws (such as the U.S. Animal Welfare Act) distinguish between "higher" animals (primates and dogs) and "lower" ones (worms and mosquitoes), which they exclude from protections. But even "higher" animals do not have the right in the United States to sue people or own property.

The legal position of animals came up in a copyright case. A crested macaque monkey named Naruto took several "selfies" of himself using David Slater's camera. Slater claimed ownership and profited from publishing the cute pictures.[6] In response, an animal rights group argued that Naruto owned the copyright and that Slater was illegally profiting from Naruto's property.

The question went to federal court. The U.S. Copyright Act of 1976 protects "original works of authorship." Moreover, for photography, the law states that the author who takes the picture owns the copyright. But what is an author? The Friends of Naruto contended that authorship under the Copyright Act is available to anyone, including an animal, who creates the original work of authorship.

The court disagreed. It cited court rulings that "if Congress and the President intended to take the extraordinary step of authorizing animals as well as people and legal entities to sue, they could, and should, have said so plainly." The judge noted that "there is no mention of animals anywhere in the Act." In the end, the court decided that no one owned the copyright on the pictures, and Slater lost his Naruto revenues.

Perhaps animals cannot sue or vote, but they have certain protections. In philosophy, the movement called *animal utilitarianism* signifies that actions should consider the happiness and misery of animals. I encounter this dilemma every summer when it is lobster time. My assignment is to cook the lobsters in a pot of steaming water. My granddaughters like to watch the lobsters run around the floor. But when it is time to put them to their death, I ask, are they suffering? How would I know since they utter no sound? If lobsters feel pain, are there less painful ways to dispatch them?

It turns out that shellfish like crabs and lobsters do experience learned avoidance.[7] If they are shocked once, they tend to avoid the shock in the future—as do mice and humans. While we cannot feel what lobsters feel, this experiment suggests that shocks like boiling water do not produce happiness for lobsters.

I will need to seek alternative methods. Switzerland has banned boiling live lobsters and requires that they be stunned before they are boiled. Perhaps I should just stop cooking lobster altogether. On the other hand, if I change to beef or swordfish, perhaps I am just outsourcing the painful business to someone else.

Animal utilitarianism poses deep difficulties. To begin with, it is not possible to respect the preferences of animals in the same way we can with humans since animals do not talk or vote. Second, it seems likely that we will give priority to different living forms—respecting dogs and chimps more than jellyfish and mosquitoes.

Which living beings have interests and which do not? One definition, held by the philosopher Peter Singer, is that sentient beings are protected while nonsentient ones are not. Sentient denotes the capability of experiencing pain or suffering, regardless of species. Therefore, dogs and lobsters as sentient are protected, while trees and sponges, which have no nervous system and therefore no feelings, are not sentient and have no individual protections.

Conservation biologists draw the line differently because they emphasize species and the tree of life. They would protect different *species* of trees or mosses as the result of the miracle of life. Does this extend to all life forms? I for one would vote for the extinction of ticks and mosquitoes, but others will provide a spirited defense here.

Finding the right balance between the needs of humans and the interests of animals is one of the most contentious subjects in all of Green fairness.

Conclusion

A leitmotif heard through this entire book is this: We cannot neatly separate Green issues from other issues in economic, social, and political life. The Green society is nested inside the broader society. Dwight Eisenhower, general turned president, put this clearly:

> Every gun that is made, every warship launched, every rocket fired signifies, in the final sense, a theft from those who hunger and are not fed, those who are cold and are not clothed. This world in arms is not spending money alone.[8]

Eisenhower's point is a subtle one about the substitutability or fungibility of resources. When we allocate our resources to one area (guns), we necessarily withdraw them from others (butter). We can compensate people who are injured in the environment with benefits in other sectors. Moreover, it may be more effective to provide people with adequate health care rather than to remove the last microgram of some harmful substance.

We can apply this fungibility principle to the case of global warming. As the chapters on global Green will discuss, nations have made only tiny efforts to slow climate change. However, even with strenuous efforts, the globe may experience 2°C or 3°C of warming in the coming century. One way of compensating those who are harmed would be to invest heavily in other areas so that the welfare of future generations will be improved in non-Green areas to offset the deterioration in Green areas. These investments will not offset the damages to everyone, such as those on low-lying islands, but they would offset the damages in the aggregate and for, say, 99 % of people in the future.

Similarly, applying the principles of Green fairness to the eating of animal-based foods is complicated because there are so many possible violations. I mentioned the problem of outsourcing pain-inducing behavior when I buy prepared food rather than put the lobster in the pot. We might therefore avoid eating any meat or fish. But do we know whether the wheat in the bread is cultivated with fertilizers that run off and cause toxic blooms in Lake Erie that strangle fish? Or perhaps our food is produced in factories that are unsafe for humans. Our hands are seldom Green and clean.

Having made a brief excursion into the issues of fairness, I conclude that, with a few exceptions, Green fairness should be viewed in the context of the broader issue of social fairness. The primary elements of unfairness in America today are malnutrition, inadequate income, poor schooling, and lack of health care. These arise in part because of the design of a fiscal system that benefits the wealthy. At the same time, some areas of Green fairness are worthy of careful attention on their own merit. Animal utilitarianism is one that is evolving over time. And including an accounting for all nonmarket impacts in government projects will help prevent some of the worst abuses of environmental justice.

PART II

Sustainability in a Perilous World

Green Economics and Concepts of Sustainability

Variants of Green Economics

What is Green economics? In one sense, it is the subject of this book. It is a growing branch of economics that deals with the environment, pollution and climate change, and the analysis and treatment of externalities. Its roots were developed by Arthur Pigou, whom we met earlier. Pigou analyzed the gap between the social and private impacts of decisions, as well as tools such as environmental or Green taxes, to close the gap or internalize the activities.

Additionally, there is a specialized field that calls itself *Green economics*. Its proponents tend to emphasize market and policy failures and express skepticism about the effectiveness of market mechanisms to produce efficient and equitable outcomes. We first introduce some of the key ideas from Green economics and then focus more closely on the key issue of sustainability.

A Vision of the Green Economy

Mainstream economics deals primarily with the workings of the market economy—health care, labor markets, and finance being some key areas. As analyzed in earlier chapters, mainstream environmental economics includes spillover effects in which market transactions have impacts outside the marketplace—to the health of humans and other life forms, to ecosystems, and to future climatic conditions.

The Green economy is a branch of economics that emphasizes the behavior of the *nonmarket systems that humans affect*.[1] An exemplary study of this area is found in a monograph by Michael Jacobs, *The Green Economy*.[2] This study has many points of commonality with the present work. However, it is highly skeptical about the ability to incorporate the environment into mainstream or "neoclassical" economics.

The mainstream view, largely adopted here, is that environmental goods and services are like normal ones except that they suffer from market failures. The remedy, in the mainstream view, is to correct the market failures and then proceed with business as usual. For example, if urban smog is the result of underpriced emissions of sulfur dioxide, then we need to price sulfur dioxide emissions properly, and the economy will then function properly.

While this view of neoclassical economics is oversimplified, it does capture the stance of mainstream economics on major environmental issues. What, in the view of Jacobs and his colleagues from Green economics, is wrong with this view? There are four major shortcomings that would need to be corrected in a truly Green economy. While I would not endorse these in their entirety, they are in the spirit of Green and need to be carefully weighed.

The first critique is that preferences (or the demands in supply and demand) do not reflect the interests of future generations. Current decisions are made by today's consumers and today's voters, and future generations have no say in these. Hence, if politicians today refuse to take steps to wreck the future oceans, future voters have no chance to vote them out of office.

A second and related shortcoming is that financial markets and public decisions do not properly weigh present and future. This bias to the present is reflected in discount rates (including market interest rates) that are too high. As is discussed in the section on behavioral biases below, too high a discount rate will overvalue present costs and undervalue future benefits. The generational tilt implies that the benefits of investments in ensuring the future health of the earth system, in preventing climate change, and in preserving precious environmental assets are undervalued. The future appears too small because of a defective telescope for viewing it.

A third major shortcoming is that mainstream economics is said to undervalue public goods such as environmental quality and environmental goods and services. These are undervalued because they are underpriced in a laissez-faire market economy. For example, certain species may become extinct because their breeding stock is underpriced and are therefore undervalued in the fish market. This applies even more powerfully for global public goods like climate change or protection of the ozone layer, where the market prices are not just low but zero. This point needs to be emphasized, but it is a key tenet of mainstream economics as well. Many prices for public goods are incorrect and indeed too low. This is seen in the fact that the price of carbon dioxide emissions in most sectors and most countries is zero and therefore well below the social costs.

A final area is that mainstream economics downplays the central concern—which in some sense encompasses the first three—which is the need to ensure *sustainability* or *sustainable growth*. Sustainability has deep roots in environmental history and has spread to economic development. We even find an "Office of Sustainability" in many organizations. What exactly is sustainability? How can we measure it? Are we on a sustainable path?

In his book *The Green Economy*, Michael Jacobs puts sustainability at the forefront of its principles. He views sustainability as about protecting the future since the interests of future generations are not represented today. He proposes two tests of sustainability to represent future interests. Here is his reasoning:[3]

Imagine we were living in a hundred years' time. What would we want previous generations to have done with respect to the environment? Two intuitive answers spring to mind. . . . A "weak" version of sustainability would require that the environment is sustained only in the sense that future generations are guaranteed the avoidance of environmental catastrophe. By contrast, the "strong" or "maximal" version of sustainability would demand rather more: that future generations are left the opportunity to experience a level of environmental consumption at least equal to that of the present generation.

One point to recognize about Jacobs's exposition of Green economics is that sustainability expresses a narrow view of human concerns since it is primarily about the environment. In the weak version, society wants to avoid environmental catastrophe, which is hardly controversial, although we would want to avoid all catastrophes, including wars and pandemics. In the maximal version, society should guarantee environmental consumption, which would appear to prioritize environmental over other items of consumption.

As will appear below, the mainstream view of sustainability takes a completely different approach—that we should ensure future generations can have an overall standard of living at least as good as the current generation. The balance of this chapter develops this third view and its implications.

Sustainable Growth: The Origins

Concerns about sustainability arose more than a century ago with writings on forestry. One idea was that forests should be managed so they provide *maximum sustainable yield*, which is the maximum timber harvest that can be sustained indefinitely.

The concept of sustainability began with forests but has been extended to other natural resources. Other sectors include nonrenewable natural resources like energy, nonfuel minerals, and soils; renewable resources like fisheries and aquifers; and vital environmental resources like clean air and water, the stock of genetic material, and our present climate.

The idea of sustainable growth was popularized in 1987 by the World Commission on Environment and Development (the Brundtland Commission):[4]

> Nature is bountiful, but it is also fragile and finely balanced. There are thresholds that cannot be crossed without endangering the basic integrity of the system. Today we are close to many of these thresholds; we must be ever mindful of the risk of endangering the survival of life on Earth.

Sustainable development was defined by the Brundtland Commission as "development that meets the needs of the present without compromising the ability of future generations to meet their own needs." It concluded that there "are environmental trends that threaten to radically alter the planet, that threaten the lives of many species upon it, including the human species."

Sustainability: The Economic Interpretation

How can we put the concept of sustainability into an economic framework? An illuminating analysis of sustainability was proposed by Robert Solow, the pioneer of economic-growth theory from the Massachusetts Institute of Technology (MIT). Solow's approach was to treat sustainability as a form of intergenerational egalitarianism, as he states here:[5]

> I will assume that a sustainable path for the national economy is one that allows every future generation the option of being as well off as its predecessors. The duty imposed by sustainability is to bequeath to posterity not any particular thing . . . but rather to endow them with whatever it takes to achieve a standard of living at least as good as our own and to look after their next generation similarly. We are not to consume humanity's capital, in the broadest sense.

In other words, sustainability means that this generation may consume its natural and produced endowments as long as future generations can also enjoy a standard of living at least as good as the current generation.[6]

The Solow sustainability criterion raises three questions: First, what are living standards? Second, what are the prospects of future generations being better off than the present? Third, what are the major threats to future well-being, and, particularly, do they come primarily from the degradation of the environment and natural resources or from other areas?

The first question involves what we are actually sustaining. The mainstream economic approach is to assume that the proper perspective is the level of consumption that individuals desire and enjoy, or what philosophers call the individualistic perspective. We should not substitute our tastes for those of the population. Rather, social conditions should be judged based on how they are ranked by members of a society.

Also, consumption should be interpreted in a broad way—it should include not only standard items such as food and shelter but also services and intangibles such as culture, leisure, and the pleasure of nature hikes. Some items of broad consumption, such as nature hikes, are omitted from conventional measures of national output because they occur outside of the marketplace. Moreover, standard measures have some important deficiencies, such as the omission of health status and many intangible investments. But items included in standard measures of national output are important and well measured, so standard metrics provide an important and objective measure of living standards.

Taking the second question, what are the prospects for economic growth over the coming decades? A starting point is to look at economic history. Economic historians estimate that global per capita real output has grown at about 2.2% per year since 1900. Until the sharp, pandemic-induced downturn in 2020, global growth over the last two decades was above the historical average.

It would require a major discontinuity for growth to turn negative for a substantial period. True, the world economy has definitely taken a hit during the COVID-19 pandemic. But expert forecasters indicate that, after what might be a prolonged downturn, the economy will eventually recover to its normal growth rate (although eventually might be many years).[7]

What are future prospects? A team of economists led by Peter Christensen used two techniques to estimate the expected economic growth rate of conventionally measured gross domestic product (GDP) over the period to 2100. One was a statistical procedure, and the second was a survey of experts. The two approaches yielded estimates of global per capita output growth of slightly above 2% per year over the 21st century. A striking feature of this study is that the two approaches, completely different in their methods, provided similar projections of future growth.[8] So the summary on the second question is that it seems likely that future generations will be better off than the present generation using standard measures of living standards.[9]

Third, how likely is a decline in future living standards? This would respond to what Jacobs called the "minimal" test for sustainability that refers to potential catastrophic downturns. The experts in the Christensen survey assessed that there is about a 5% probability that the growth rate to 2100 will be negative—in other words, that people living in 2100 will be worse off than those living in 2010. The statistical technique projected an even lower chance of economic decline.

The survey also asked experts to identify the threats to future economic growth. Four respondents believed that wars would be the largest threat, while one believed that catastrophic climate change would be the cause. Surprisingly, not a single one of the experts mentioned pandemics as a major threat to the future economy.

So, on the third question, both statistical techniques and experts find that the chances of economic decline during this century are very slim. But experts cannot accurately predict the known unknowns and can hardly be expected to foresee the unknown unknowns, so we must take these projections with caution.

Components of Sustainability

The major difference between Green economics and mainstream economics concerns the application of the sustainability concept. Green economics focuses on the central importance of environmental consumption, while mainstream economics assumes that a broad bundle of goods and services—nonenvironmental as well as environmental—is the goal of economic activity.

To begin with, mainstream economics assesses the sustainability of a *broad* range of assets and a *rich* array of consumption goods and services. This approach allows the substitution of more abundant assets and goods for those becoming scarcer. Robert Solow put the point this way:[10]

> It makes perfectly good sense to insist that certain unique and irreplaceable assets should be preserved for their own sake, Yosemite [for example]. But that sort of situation cannot be universalized: it would be neither possible nor desirable to "leave the world as we found it" in every particular. Most routine natural resources are desirable for what they do, not for what they are. It is their capacity to provide usable goods and services that we value.

The tendency of consumers to find less expensive ways of satisfying their needs is the fundamental principle of *substitution*. Substitution occurs when needs are met by substituting goods that have declining prices and higher quality for those with rising prices and stagnant quality. Economic history is a book with many chapters on new technologies that led to the substitution of new, higher quality, and less expensive goods and services. There are chapters describing air travel replacing trains, which in turn outperformed stagecoaches, toilets ousting outhouses, cell phones substituting for landlines, and emails outpacing postal letters. We can reasonably ask whether the principle of consumption substitution applies everywhere. Are some elements of consumption sacred and inviolable?

We see no clear answers here, and indeed the answers are evolving over time. Most people would agree that societies should protect certain unique and irreplaceable assets (like Yosemite) as well as religious or cultural items (such as sacred temples). In the United States, free speech, the right to a trial, and the right to vote are inviolable principles, at least in principle. We cannot sell ourselves into slavery, even when we are in the most desperate situation. No one but the most extreme market fundamentalist would want to auction

Yosemite for mining development or sell New York's Central Park for a city of Trump Towers.

But other items are not inviolable. For conceptual clarity, let us call goods without sacred or inviolable elements *pure economic goods.* The major point, as Solow explained, is that sustainability does not require preserving pure economic goods for future generations. Prior generations had no obligation to this generation to maintain a minimum supply of outhouses or stagecoaches or kerosene lamps when cheaper and more desirable substitutes became available.

Similarly, we have no obligation to future generations to provide a minimum quantity of toilets or automobiles or bulky laptop computers. Sustainability requires adequate food, shelter, and health care. It does not require that houses be built from trees rather than synthetic materials, or that we eat wild rather than farmed fish, or that we live in small houses and drive big cars rather than live in large houses and drive small cars.

However, the stance of Green economics as represented by Jacobs is that certain environmental activities and assets are inviolable rather than pure economic goods. It is not, in that view, acceptable to provide a lower quality of environmental services so that people can enjoy a greater amount of nonenvironmental goods and services. For example, a biocentric viewpoint might hold the existence of major species to be beyond economic trade-offs. Or perhaps the existence and future enjoyment of pristine forests should not be sacrificed for normal goods.

Is there a role for red lines, for inviolable standards, here? And if so, where is the line? I would respond that we need to be cautious in drawing reds lines for social decisions and elevating some activities to the status of absolute necessities. We should always ask whether the environmental goals are valued for what they do or what they are.

Here are some areas where there is a lively debate about where to draw the red line. Two important areas are species survival and preventing climate change. I would argue that societies cannot escape from weighing costs and benefits, even if we would like to draw red lines to simplify decisions. Similarly, there is no bright line on how much pollution to allow or where the boundaries of protected lands should be placed. The dilemma in a pandemic—how much to shut down society to reduce infections versus open up to reduce unemployment—is an unavoidable choice. For those situations, the ethical dilemmas we face generate fierce and genuine differences of approaches that cannot be finally resolved by religion, environmentalism, science, or economics.

A Parting Vision

We cannot end a discussion of sustainability without asking, sustainability for what and for whom? For this, we turn to Columbia University's Jeffrey Sachs. More than any single person today, Sachs has been a brilliant and tireless scholar-activist for sustainable development informed by the best economic and environmental thinking. He summarizes his vision as follows:[11]

> The fact of the matter is that humanity is still rushing headlong towards multiple collisions with nature and with each other, within highly divided and unfair societies. And yet, we have the means to succeed; that is, to combine the end of poverty with social inclusion and environmental safety. The most essential quality for our survival will be a shared moral impulse to do the right thing: to protect each other and nature from our greed, scientific lack of understanding, and moral disregard and carelessness.

Sachs's summary of sustainable development, and his warning about collisions with nature, parallels the conclusions of this book as well.

Green National Accounting

I remember the exact moment I became interested in Green accounting. I was flying out of Albuquerque, reading a glossy magazine on a defunct airline, TWA. I saw an article criticizing gross national product (GNP) and encountered the following: "In the words of a young radical, don't tell me about your GNP. To me, it's really Gross National Pollution."[1]

I thought, wow, that's really cute. But is it true?

Actually, it is completely false. Our output measures do not count pollution. They include goods like cars and services like concerts but not carbon monoxide (CO) pumped into the air.

However, the complaint has a subtle point worth considering— the measures of national output do not adequately correct for pollution or other spillover effects of the economy. A set of accounts that properly deals with pollution is called *Green output*. We will see that a serious effort has been made to develop these accounts but that this is extremely difficult terrain.

How Do We Measure National Output?

I pause for some background on how we measure output. Most discussions of national output talk about gross domestic product, or GDP. GDP is the value of the goods and services produced by the nation's economy less the value of the goods and services used up in production. Hence, it includes consumption goods like food and investment goods like new houses, as well as production for government and adjustments for foreign trade.

In 2018 the per capita GDP of the United States was $62,600, which was the highest of large countries. That of the most populous country, China, was $18,200. The poorest large country was the Democratic Republic of the Congo, with a per capita GDP of around $930. There are many subtle difficulties in calculating these numbers, but they are the best we have at present. Here is the way a leading economics textbook explains the importance of this measure:[2]

> Of all the concepts in macroeconomics, the single most important is national income and output, particularly GDP. While the GDP and the rest of the national accounts may seem to be arcane concepts, they are truly among the great inventions of the twentieth century. Much as a satellite in space can survey the weather across an entire continent, so can the GDP give an overall picture of the state of the economy.

While it is universally measured and used, GDP has its critics. One elementary problem is that GDP includes gross investment and does not subtract depreciation. Hence, it includes all new houses built in a year but does not subtract the houses that are burned up by wildfires. Because it does not subtract depreciation, gross investment is too large a number—too gross.

A better measure would include only net investment as part of total output. Net investment equals gross investment minus depreciation. It is also useful to focus on the income of residents, which would be represented by national product rather than domestic product. By subtracting depreciation from GDP and looking at the income of residents, we obtain net national product (NNP). If NNP is a sounder measure of a nation's output than GDP, why do national accountants focus on GDP? They do so in part because depreciation is difficult to estimate, whereas gross investment can be estimated fairly accurately. Additionally, GDP is familiar, and statisticians are reluctant to change a concept that is so widely used.

But even NNP has its limitations. While it includes all the goods and services produced by residents of the country, it excludes those that are not produced and sold in markets. Therefore, it includes the lumber from forests but not the value of the forests' nature hikes or erosion control. It includes the electricity produced and sold by an electric utility but not the health damages from the pollution that utility emits. So, while the young radical was wrong to claim that GDP includes pollution, a correct statement would be that GDP and NNP do not include a subtraction for pollution.

> So here is the first definition. *Green output* is a measure of national output that includes important nonmarket goods, services, and investments along with corrections for the impacts on the economy of externalities such as pollution.

Weitzman's Brilliant Theory of Environmental Accounting

Most specialists would agree that it is important to correct for pollution, climate change, and other nonmarket activities and externalities in the economic accounts. But how is this done in practice? How could we figure out how to subtract the economic harm done by water pollution or carbon dioxide (CO_2) emissions from the value of food and shelter?

This seems an impossible question. But a striking analysis by Harvard's Martin Weitzman (1942–2019) showed the way.[3] Weitzman's approach, which has been incorporated in full-income accounting, or Green accounting, is actually quite intuitive. The idea is to extend the standard national economic accounts—which cover market transactions—to include nonmarket activities or processes. The approach of the standard accounts is to collect data on the quantity of production and the prices (of apples, lumber, gasoline, cars . . .), calculate the values as the product of prices and quantities, and then calculate total national output as the sum of the values of final outputs sold to consumers and other sectors.

The standard accounts are indeed flawed, but not in the manner of the young radical quoted above. The problem is not that pollution is included in the standard accounts—it definitely is not. Rather, the problem is that pollution is *not* included in output when it actually should be included. The Weitzman approach assumes that the harmful externalities are priced and then adds the value of the externalities to the totals. So Green NNP = normal NNP + the price times quantity of pollution.

Is it all that simple? The tricky aspect is to remember that harmful activities have a *negative* price because they are "bads" rather than "goods." Therefore, the price times quantity of pollution would be a *subtraction* from national output rather than an addition. Hence, for example, if there are 5 million tons of air pollution in a year and the damage from air pollution is $100 per ton, this would require a subtraction of $500 million from national output.

All this is straightforward—except that the concept of the "price of pollution" may be puzzling. The price of potatoes is observable in the grocery store. That is the price that the grocery store charges, and it is also the cost to consumers. But what is the price of pollution (perhaps CO from a truck)? From the point of view of the firm and its commercial accounts, the price is zero. And that is why there is no item called "sales of air pollution of carbon monoxide" in the national economic accounts. But the cost to people is not zero because pollution does damage to human health. To return to the example in the last paragraph, perhaps each ton of CO emitted does $100 of damage. According to the Weitzman approach, the damage is the appropriate price to use when subtracting the costs of pollution and other externalities in calculating full national income or Green output.

Is the problem thereby solved? In principle, yes. But in practice, calculating the costs of pollution and other externalities is extremely difficult because the data are sparse at best and missing at worst. This point was well made by a committee of the National Academy of Sciences in the following passage:[4]

> Consider the problems involved in accounting for a simple loaf of bread [using the Weitzman technique]. Doing so would require

measuring and valuing a wide variety of flows of water, fertilizer, pesticides, labor, climate, and capital inputs that go into producing the wheat . . . the complex combination of human skills, equipment, and structures that go into milling the wheat [and so forth]. It appears unlikely that anyone would try, and safe to conclude that no one could succeed in, describing the physical flows involved in this little loaf of bread. Fortunately, however, [standard] economic accounting does not attempt such a Herculean task. Rather, the national accounts measure all these activities by the common measuring rod of dollars. . . . The above comparison may give some sense of why accounting for environmental flows outside the marketplace is such a daunting task.

So here is where we currently stand: We have comprehensive accounts on the market economy for most countries. We can calculate the standard concepts like GDP or NNP, as is suggested in the last quote, because we can use easily observed magnitudes of dollar flows and prices.

By contrast, we have only minimal information on accounting for externalities because there are very sparse data to construct the prices and quantities of nonmarket activities. Scholars have been working on this problem for almost half a century, but we still know relatively little. The next section links standard measures of national output with the concept of sustainable output while the following section provides some illustrative estimates of how current estimates can be extended toward a more comprehensive Green output.

Net Output and Sustainable Output

Green national output provides an important and surprising link between standard economic measures and the concept of sustainable output. As we saw in the last chapter, the economic definition of sustainable output is a level of consumption that allows future generations to be at least as well off as current generations. We further saw that sustainable output has its roots in forestry with the concept of sustainable yield. The sustainable yield of a forest is the amount that can be harvested indefinitely. An alternative definition, which is closer to economics, is the maximum harvest that will leave the forest stock intact and therefore able to produce the same harvest in the future.

Starting with this forest perspective, we can introduce the more general economic definition of sustainable output. This would be the maximum that an economy can consume while leaving the same capital stock for the next year or the next generation.

The concept of sustainable output is illustrated by a fruit-tree economy. Suppose there are 1,000 trees producing 100 fruits, which can either be eaten or planted to grow more fruit trees. We will construct national accounts for the economy, measuring output in fruit units, so the output is 100F. Each year 10 trees die. We therefore need to set aside 10 fruits to grow replacement trees. That leaves ninety fruits, which can be consumed each year while leaving tree capital (the number of trees) intact. Therefore, the gross output of the economy is 100F while the net output is 90F.

We can extend this by supposing that the economy is expanding and adding 10 trees each year. So consumption is 80F while net investment is 10F. Net output and sustainable output (consumption plus net investment) are still 90F. In this simple example, net output (90F) equals gross output (100F) less depreciation (10F). It also equals consumption (80F) plus net investment (10F). The important point here is that properly measured net output (90F) also equals the maximum sustainable consumption, which is also the same as sustainable output.

The fruit-tree example can be extended to a more complex economy with many goods, services, and types of capital. But the basic proposition holds in the more complex system as well. *In an economy where all inputs and outputs are properly measured, sustainable output can be calculated as net national product, or consumption plus net investment.* This important result suggests why measuring Green output should be at the top of the research agenda for Green economists. This agenda would include the kinds of corrections that now follow on excluded and mismeasured activities.

Selected Corrections for Excluded Environmental Activities

I emphasize that no comprehensive environmental accounts exist for any country—indeed, there are only the scantiest of accounts. However, we can use the sparse existing research to get a flavor of such accounts. This discussion focuses on three sectors where environmental accounts have been or could easily be constructed. These are greenhouse-gas emissions affecting climate change, sub-soil minerals, and air pollution.

From a conceptual vantage point, the starting point is net national product. In developing these estimates, we can calculate both a *level correction* and a *growth correction*. The level correction adds or subtracts the estimates of the externalities or other omissions from NNP.

So perhaps the correction for pollution X is 1.0% of NNP in 2014 and 1.1% of NNP for 2015. The growth correction looks at the impact of the correction on the growth of NNP. If the externality is growing, then this will reduce the growth rate, while if the externality is shrinking, this would increase the growth rate. Using the numbers just given, the pollution correction would lower the growth rate from a conventional growth rate of NNP of, say, 3.0% to a corrected rate of Green NNP of 2.9%.

CLIMATE CHANGE

Let us now turn to some actual cases. The first example is the impact of the climate-change externality, particularly CO_2. Unlike the next two examples, this one is so simple that anyone can calculate it on a spreadsheet. The idea here is to obtain estimates of the quantity and the price and then correct the accounts for the total. You would begin with a measure of greenhouse-gas emissions, here for CO_2. You then multiply the quantity by the price of emissions. For the price, we use the *social cost of carbon* estimated and used by the U.S. government (see the discussion in the chapter on Green politics).

TABLE 9-1. Calculation of environmental correction for climate change

Year	Official NNP [billions of 2012$]	CO_2 emissions [millions of tons of CO_2]	Price of CO_2 [$/ton of CO2, 2012$]	CO_2 correction [billions of 2012$]	Corrected NNP [billions of 2012$]
1973	5,227	4,735	11	53	5,043
2018	15,872	5,317	43	229	15,699
Annual average growth rate, 1973–2018	2.468%	0.257%			2.493%

Source: The estimates in table 9-1 calculate real output using a Törnqvist index. Data for CO_2 emissions are from the U.S. Energy Information Administration, output data are from the U.S. Bureau of Economic Analysis, and the social cost of carbon (SCC) is from the U.S. Environmental Protection Agency. For the period 1973–2015, the SCC is assumed to grow at 2% per year in real terms. The estimates of the SCC for the 2°C target are taken from results using the DICE model as reported in William Nordhaus, "Climate Change: The Ultimate Challenge for Economics," *American Economic Review* 109, no. 6 (2019): 1991–2014, doi:10.1257/aer.109.6.1991.

Table 9-1 shows the calculations. For these we use constant prices. Focus first on the line for 2018. Here, total CO_2 emissions for the United States in 2018 were 5.3 billion tons. The U.S. government estimates that the social cost of emissions for 2018 was $44 per ton. Therefore, the total subtraction is $44 × 5.3 = $229 billion. This would be a debit from the $15,872 billion of output in that year, or a level correction of 1.5% of output.[5]

Next, calculate the growth effect. For this calculation we begin with the corrected NNP shown in table 9-1 for the years 1973 and 2018. We see that the CO_2 correction grew slowly over the period— reflecting that emissions declined 2.2% per year relative to output. The growth effect of the climate correction was counterintuitively slightly *negative*. So Green NNP rose faster than conventional NNP. To be precise, after correction, the output growth was 2.493% per year for the 1973–2018 period using the corrected figures instead of 2.468% per year using the official figures. The negative growth effect is counterintuitive until we realize that it arises because CO_2 emissions declined, so their effect on Green output was larger at the beginning than at the end. The growth effect was small (negative 0.025% per year) but still a surprise. Thus, correcting for CO_2

emissions would lower the *level* estimate of output but would raise the *growth* rate of output by a tiny amount.

One wrinkle here would be to ask what the growth correction would be with more ambitious climate targets. As we will discuss in the chapters on global Green, international policy has a target of a 2°C limit on temperature increase. This would imply a much higher social cost of carbon and therefore a much higher price of carbon in the calculation in table 9-1. One estimate for this target is that the carbon price would be more than 5 times larger with the more stringent target. Using the same method as shown in table 9-1, the level correction for the 2°C target is much larger, at 8% for 2018, and the growth correction is also correspondingly larger. When environmental costs are larger, this implies that true output is also lower than conventionally measured output. But when environmental impacts are declining, the growth correction is positive but is also larger.

SUBSOIL ASSETS

The second sector of interest for Green national output, more complex than the first but reasonably manageable, is subsoil assets. These include stocks of oil, gas, gold, silver, copper, and other metals.

What is the flaw in their standard treatment? The issue is that subsoil assets are not properly calculated in national output because there is no accounting for depletion and additions. Subsoil assets are, in effect, valuable ripe fruits hanging on trees that are ready for harvesting. We do not deduct the value of the oil-in-the-ground (fruit-on-the-tree) when we harvest these assets (this being depletion). We also do not add the discovery of new reserves (the growth of the fruit-on-the-tree), these being additions.

The most careful study of the impact of omitted mineral depletions and additions was by the Bureau of Economic Analysis and dates from the 1990s. The result was that depletions and additions were each on the order of .5% of NNP, and the net level and growth effects were both zero. The reason for the minimal impact is that both quantities and dollars of additions were close to quantities and dollars of depletions. Looking at oil and gas with more recent

data, we find that, quantitatively, additions are larger than depletions (reserves of both oil and gas are increasing). We cannot be
sure that the dollar values are the same as the quantities because the
additions might be low grade and less valuable than the depletions,
but this is purely speculative. Therefore, this second component of
environmental accounts suggests that the impact of correcting for
subsoil assets is close to zero.[6]

AIR POLLUTION

The third and most complicated important example is air pollution.
This includes some of the deadliest and costliest externalities, those
associated particularly with burning coal and other activities. Most
of these are regulated in the United States, but few are priced at a
level that reflects their social costs.

I will report on one study that calculated environmental accounts
for air pollution performed by Muller, Mendelsohn, and Nordhaus,
with updates by Muller.[7] This study estimated air pollution damages in the standard manner described above. That is, total damages
were calculated as the price (damages per unit pollution) times the
quantities of five major pollutants (nitrogen oxides, sulfur dioxide,
fine particulate matter, ammonia, and volatile organic compounds)
for ten thousand sources. The volume of emissions was calculated
at each source location for each industry, and the damages were
estimated for each of the U.S. counties.

The major corrections of the accounts were for industries like
coal-fired power plants and stone quarrying. Total damages as a
percent of NNP declined from 6.9% of output in 1999 to 3.4% of
output in 2008. These corrections are clearly a substantial fraction
of output—and are also a much larger fraction of the output of the
highly polluting industries.

The growth effect was again counterintuitively *negative.* The reason is that the pollution subtraction at the end of the period was
smaller than the subtraction at the beginning (as with the case of CO_2
discussed above). The growth effect of pollution was to raise total
NNP growth from 2.03% per year to 2.45% per year—a substantial

impact that has not been emphasized in discussions of the economics of pollution.

The three examples pursued here do not exhaust the areas of interest. Other Green sectors would include forests, water, congestion, and toxic wastes, but there are few estimates for those. Estimates of augmented accounts have been produced in other areas such as health, home cooking, family care, and leisure. These can have substantial effects on total output and on growth, but they are generally outside of the purview of Green accounting.

Summary Judgment on Green Accounting

Here is the summary of Green national output: When we include estimates of the impacts of resources and the environment that are currently excluded from the conventional national accounts, these can make a substantial difference in the *level* of output. A rough estimate is that the impacts of excluded sectors such as those reviewed here would subtract on the order of 10% of output from the United States, but since the research is incomplete, the total might be larger.

However, and paradoxically, correcting this omission will tend to raise the *growth rate* of Green output, at least for the last half century for the United States. The reason is that most measures of pollution have been declining relative to the overall economy—the result of cleaner power plants, factories, and automobiles. It is the growth of pollution relative to other goods and services that affects the growth rate. The growth effect in the sectors examined to date is on the order of *plus* one-half percentage point per year—a substantial number that would add up considerably over the years. True, major sectors are missing from the estimates. But, while approximate, these numbers do cover some of the most important externalities.

The finding that environmental policies are adding to genuine economic growth is important for debates about environmental policy. I would count this as a major victory for the Green movement. The reason for this surprising finding is interesting. If we go back half a century, to the dawn of environmental regulation in the United States, externalities such as air pollution were activities in which the marginal benefits of reducing pollution were far greater than the marginal costs. Therefore, environmental policy was, in effect, picking low-hanging and inexpensive fruit, reducing health and other damages substantially at minimal cost.

If we look only at the standard economic accounts, we will largely miss the improvements in economic welfare associated with picking the low-hanging environmental fruit because the health benefits of environmental regulation are not counted in the standard accounts. However, if we extend our horizon to include external benefits, the environmental policies have actually improved growth substantially.

So if the young radical were to come back today as an old radical, the attitude toward national accounts might be quite different. Having seen the experience of recent years and studied the work of environmental economists, the old radical might write, "Those who claim that environmental regulations harm economic growth are completely wrong because they are using the wrong yardstick. Pollution *should* be in our measures of output, but with a negative sign. If we use Green national output as our standard, then environmental and safety regulations have increased true economic growth substantially in recent years."

10

The Lure of Exo-civilizations

A central scientific and economic question for the future is whether human civilization on Earth is unique or whether it can be replicated in space or on other planets in what I will call *exo-civilizations*. Most science fiction and popular culture assume that, yes, Earth can be replicated. They assume that we can set up colonies on the moon, or Mars, or some distant planet in much the way the Pilgrims established settlements in Massachusetts. Perhaps life would be harsh at first but after adapting to the new environment, a sustainable civilization would be possible on another planet.

In reality, the question of whether human civilization can be sustained outside our planet is a deep and unanswered question. Begin by considering what is being replicated. Earth is a vast natural and human ecosystem of natural and produced assets. The Earth's resources include its oceans and rivers, oxygen, fossil fuels, rare earth minerals, and biological species. To these are added the critical ingredients of human intelligence, labor, and specialized skills, including a rich array of produced capital such as domesticated animals, cities, roads, houses, machines, factories, defenses, and the technologies developed by humans.

Finally, these are organized through institutions such as laws, governments, collective activities, and markets. These human and natural systems did not arise spontaneously but evolved to cope with the challenges of sustaining billions of humans and countless other life forms on Earth. Life on Earth is not adapted to the challenges of Mars or other planets.

If we focus on humans, Earth's system is today enormously productive, with an annual net output of goods and services of around $100 trillion (or about $15,000 per capita) being the fruits of this complex system.

Is it possible to build a closed or nearly closed system to replace or parallel earth systems? Is it possible to have a system producing not only the food and energy but also the houses, gardens, nature walks, ski slopes, sushi, baseball games, and other necessities and amenities of modern life? Perhaps we cannot replace this item by item, but we might instead have a comparable menu of Alpha Centauri cuisine, rockball games, sand resorts, volcanic walks, and other substitutes.

Where can we look to help understand the prospects of sustainability outside Earth? This chapter surveys the landscape here. It begins with tracing the long, winding, and tortuous path that led to our current human civilization. It then reviews speculations about life on other planets from the insights of current space missions.

The final section examines the question by looking at the ultimate Green Dream—the largely forgotten glass bubble in Arizona called Biosphere 2. This bold experiment attempted to establish a closed, self-sustaining system and has important lessons for the broader questions raised in these chapters.

These three stories contain a central message—that it was historically extraordinarily difficult to achieve a sustainable civilization on Earth. It will be an even greater challenge to establish a self-sustaining system elsewhere, on other planets.

The Miracles of Human Civilizations

One way to consider the prospects of exo-civilizations is to remember how long it took to arrive at our current prosperous world, even with all the advantages of human brains and Earth's bounty. The emergence of the modern world was slower than glacial. The first step, taking perhaps 4 billion years from the first stirrings of bacteriological soup, was the evolution 50,000 years ago of anatomically modern *Homo sapiens*.

The economies of early humans differed little from other animals scratching subsistence off the land and the sea. The evolution of civilizations—the long and meandering development of tools and technologies—is usefully divided into two stages: the first begins with the earliest humans and stretches to the Industrial Revolution around 1750, and the second covers the time up to the present.

The first stage, actually a crawl, involved developing the most rudimentary elements of technology: the taming of fire and animals, the invention of the stone ax, farming, the development of written languages, and the clustering into cities. It appears that each of these developed independently in different isolated parts of the world, so they were within the capability of modern humans but not of other species.

Reconstructing Economic Growth

Reconstructions suggest extraordinarily slow growth in living standards in the early human period. According to the best estimates of economic historians Angus Maddison and Brad DeLong, the growth in output per capita from the earliest times to the mid-eighteenth century was a factor of 2, or an annual average growth rate of 0.001% per year. We can think of the first 49,700 of the 50,000 years of modern humans as a Malthusian period in which improvements in technologies led to the multiplication and spread of humans (say, to the colder regions with the help of fire), rather than an upgrading of their economic status.[1]

Table 10-1 shows the best reconstruction of the economic history of humans. Populations were living at subsistence levels in the earliest periods. Reasonably accurate data are available for Roman and Byzantine times, for Western Europe from about 1750, and for most of the world in the last half century.

The message from table 10-1 is that virtually no growth occurred in output per capita and living standards for most of the human period.

TABLE 10-1. Growth of population and living standards since the earliest humans

	Per capita output		Population
Period	Level (2011 $)	Growth from prior period (% per year)	Growth from prior period (% per year)
1 million BCE	551		
0 CE	655	0.00002%	0.00034%
1000 CE	801	0.020%	0.002%
1750 CE	1,074	0.04%	0.06%
1900 CE	2,048	0.43%	0.21%
1980 CE	7,352	1.60%	0.54%
2017 CE	15,317	1.98%	0.62%

Source: see the footnote in the text.

The revolution in living standards gathered steam after 1750, then took off in the twentieth century. Global per capita output is today perhaps 30 times the level in the early Malthusian period. The story of the Industrial Revolution has been a staple of economic historians for many years. It involved the fruits of the scientific revolution of the prior period, the growth of regional and international trade, the routinization of innovation, the exploitation of the necessary resources and raw materials, the development of large companies and their economies of scale, and above all a phenomenal cascade of new technologies.

THE EVOLUTION OF LIGHTING

Familiar measures of living standards and productivity, such as the gross domestic product (GDP), are some of the great inventions of the twentieth century. They are, however, severely limited in their historical range. The United States has official output data back to 1929 and reasonably accurate data back to the 1880s. China's output data are only modestly accurate, and even rudimentary accounts for China are unavailable before 1950. Output measures for much of tropical Africa are still unreliable. It is therefore difficult to attempt to measure productivity in the distant past, particularly before the

Industrial Revolution. The data shown in table 10-1 are the best we have but are highly speculative for the early periods.[2]

An alternative approach to measuring productivity focuses on a narrow but well-measured sector, lighting. Productivity data here are the longest available and measure technological changes in illumination since the earliest times of human history. Key milestones were the taming of fire (at least 600,000 years ago), early open lamps (30,000 years ago), candles (perhaps 5000 years ago), closed lamps (from early Greece of about 4000 years ago), and recognizably modern oil lamps (from 1782). Revolutionary changes in both devices and energy forms over the last two centuries led to continued rapid improvements in lighting productivity; these involved kerosene and electricity as energy and incandescent to fluorescent to, finally, LED bulbs.

We can measure the price and efficiency of light for each of these technologies along with the hourly wage rate to get a rough estimate of lighting productivity. The price of light divided by the wage is an estimate of how much light an hour of work will buy. It is measured as lumen-hours per hour of work. Output per hour is a simple but reliable estimate of productivity.

What does this measure of productivity show? The output measure here is 1000 lumen-hours. This is approximately what a conventional 100-watt incandescent bulb would yield over an hour. The first reasonably accurate estimates date from Babylonia around 1750 BCE (before the common era). A rough estimate is that it took about 40 hours of Babylonian work to buy enough oil to produce 1000 lumen-hours of work. Gradual improvements over the next 3,500 years reduced this to about 5 hours of work. Then, with the revolution in lighting, the time cost of light declined rapidly. With today's LED light bulb, the cost declined to about 0.000072 hours per 1,000 lumen hours. Light went from being precious to essentially free.

Figure 10-1 shows the best reconstruction from 1750 BCE through 2020. It is a ratio scale, so the slope equals the growth rate, with the numerical growth rates shown for the two major periods (before and after 1800). The break in trend in 1800, visible and striking, validates the estimates for total productivity in table 10-1.

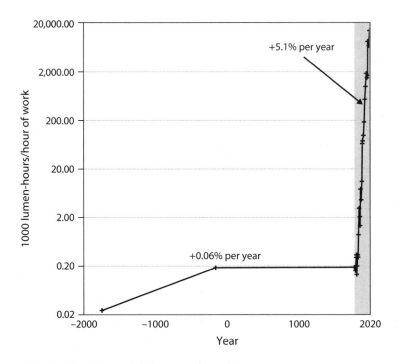

FIGURE 10-1. Productivity in lighting over four millennia
This graph shows labor productivity in illumination. Since it is a ratio graph, the slope is the growth rate, with the average growth rate shown for the two subperiods.

Table 10-2 shows the numbers by subperiod, along with the major technological developments. The two periods with the greatest improvements were around 1900, after the development of electricity, and the decades that have ushered in new technologies such as LED lighting since 1990.

The key point here emphasizes that human history witnessed a sharp inflection point occurring with the Industrial Revolution around 1750—this being the second stage of human civilizations after the development of foundational inventions like the wheel. Productivity in lighting was less than 0.1% per year from Babylonian times until the Industrial Revolution, then accelerated to over 5% per year since that time.

We also note that the productivity revolution in lighting was extraordinarily Green in its implications. One happy environmental

TABLE 10-2. Productivity growth of lighting in different eras

Start year	End year	Rate of productivity growth (per year)	Technological change (from start year to end year)
−500,000	−20,000	0.00003%	Neolithic lamps
−20,000	−1750	0.00102%	Babylonian lamps
−1750	−150	0.13%	Roman lamps
−150	1800	0.00%	Candles
1800	1850	1.17%	Whale oil lamps
1850	1900	5.22%	Town gas
1900	1950	9.53%	Edison light bulb
1950	1990	2.86%	Productivity in electricity
1990	2005	9.38%	Compact fluorescent
2005	2018	5.49%	Light-emitting diodes (LED)

effect of these new technologies, as Louis Stotz remarked, was that "the discovery of petroleum in Pennsylvania gave kerosene to the world, and life to the few remaining whales."[3]

I recount the parallel histories of overall productivity and lighting because they emphasize the long road to the affluence of the modern world. Modern humans evolved after a long and tortuous evolutionary journey of billions of years of life. However, even the presence of anatomically modern humans did not guarantee high productivity anywhere on the globe. Rather, for the first 99+ % of human history, productivity crept forward at a snail's pace.

Given the glacial pace of technological improvements over the 50+ millennia of human history, can the economic prosperity of today's Earth economy be replicated on a sustainable basis elsewhere? The history of human civilization indicates how high the barriers were to building a viable enterprise on Earth. Even today, with modern technologies, some regions are living in conditions not far above those of our Stone Age ancestors. To replicate in a short time on a distant planet what took so long to build on Earth looks to be an enormously difficult mission outside the favorable cultural, economic, scientific, and resource environment of today's modern societies.

Exo-civilizations: Life on Mars and Beyond

The challenges of constructing a sustainable society look daunting when we trace the history of human civilizations. Another perspective would be to imagine humans colonizing other planets—what I call *exo-civilizations*. Perhaps, we might think, it would be like the Pilgrims setting off to discover the New World. Settling in the Americas was risky and dangerous. But the Europeans eventually succeeded in populating a rich and powerful continent.

However, a closer look suggests that the Pilgrims are a poor analogy for the prospects of an exo-civilization. The most promising place to start a new civilization is Mars. It is nearby (by astronomical standards), has a few earthlike qualities, and is well studied. One of the proponents of remote colonization is the technologist-entrepreneur Elon Musk. Here is his vision: "I'm talking about sending ultimately tens of thousands, eventually millions of people to Mars." His plans go well beyond the red planet: "We'll go to the moons of Jupiter, at least some of the outer ones for sure, and probably Titan on Saturn, and the asteroids. Once we have that forcing function, and an Earth-to-Mars economy, we'll cover the whole Solar System."[4]

Musk is sober about the costs of colonization: "My rough guess is that for a half-million dollars [per person], there are enough people that could afford to go and would want to go [to Mars]. But it's not going to be a vacation jaunt. It's going to be saving up all your money and selling all your stuff, like when people moved to the early American colonies."[5]

The idea of space tourism seems possible in the coming decades. But is it likely that we will be able to establish self-sustaining civilizations "like when people moved to the early American colonies"? While not impossible, the prospect seems extremely remote because of both the costs and the dangers of planetary colonization.

While many of us have enjoyed science fiction writings and movies, we need to dig deeper to find serious analyses of space colonization. This section is deeply informed by a recent book by Adam Morton on space colonization, as well as a technical analysis by Sydney Do and others.[6] These studies raise two major issues: costs and dangers.

The first issue concerns whether a space colony would be self-sustaining. Relying on the last chapter, I define sustainability as the ability to have a system (or economy) that produces a reasonable standard of living while keeping its capital intact (or replacing any used-up capital). This would require producing food, shelter, health care, transportation, and energy or exporting sufficient quantities of Martian goods to pay for imports of needed goods from other planets (presumably, from Earth).

Begin with costs. An analogy to space colonies would be the costs of maintaining humans on Antarctica. Forbidding as that is, Antarctica is actually a congenial place: warmer than Mars with many superior features, such as an atmosphere, plenty of frozen water, and easy transport to the rest of the planet. The cost per scientist is about $200,000 per year, and this serves as a useful lower bound on the costs of further destinations.

Another comparison, which gets closer to space, is the International Space Station (ISS). This is a habitable satellite in low Earth orbit. It has been continuously occupied since 1998, with more than 240 occupants. According to Morton, the ISS cost $150 billion to build through 2010. A rough calculation shows that the annualized cost was roughly $600 million per person.

For an economic analysis of non-Earth civilizations, consider the Mars One program. Mars One is a private European company with plans to establish a permanent human colony on Mars. It would send four people at a time on one-way journeys to establish the Martian colony. This mission is similar to that publicized by Elon Musk. (For prospective tourists, note the emphasis on "one-way" tickets.)

A thorough analysis of Mars One by Do et al. concluded that, as planned, it is "infeasible." Many of the proposed technologies, such as the food supply and supply chains, do not currently exist. Even so, they estimate the costs of establishing the colony to be extremely high. By the time 40 colonists were in place, the cumulative launch costs would be more than $100 billion, or $2.5 billion per person. This excludes the costs of habitation, local production, communication, transportation, or spare parts. If the annual cost per person is a conservative $250 million per year, it is hard to imagine any exports that would pay for a small fraction of this cost. So the colony would not pass the sustainability test.

We might imagine that launch and other costs decline. But there are even more daunting obstacles to exo-colonization. Many of the dangers are physical. Ultraviolet radiation on Mars is much stronger than on Earth, solar energy and gravity are much weaker, and the level of light is low. There are ferocious dust storms. And it is very cold, with temperatures as low as −125°F. Additionally, there is no protection against asteroids because Mars has no atmosphere. Recent estimates indicate that about 200 asteroids hit Mars every year, and these will destroy any people, structure, or equipment in their path.

Perhaps all these perils can be overcome with sufficient investments and ingenuity. But there remain questions of psychological, economic, and social structures. Take the simple question of pets. Americans have almost 100 million dogs. They provide companionship and love. They are also valued professionals as guides, as herders, in search and rescue, in therapy, in detection, and as soldiers. But, like humans, dogs have evolved and adapted to the special environment of Earth and man. They are unlikely to find a home on the dangerous soil of Mars, so it would be a lonely place. The exo-colonialists would also miss other products like fish, tomatoes, milk, cheese, and meats. Moreover, AmazonMars would require almost a year to deliver your orders in the swiftest of spaceships.

We cannot render a verdict on the future. But the prospect of a self-sustaining exo-civilization on Mars or elsewhere outside of Earth seems remote. Not impossible, but surely infeasible with anything resembling today's technologies.

Biosphere 2 as a Laboratory for Sustainability

A final step in our examination of sustainability is perhaps the most instructive. This was an experiment to test the possibility of setting up a closed system here on Earth—Biosphere 2.

Biosphere 1 is Earth itself. So what was Biosphere 2? This was a private venture designed to prove the viability of a closed ecological system. The mission was to show that eight humans (*biospherians*) could produce enough food to live for two years without any external food supply. Note at the outset that the mission was not really aimed at sustainability. Food is only a tiny fraction of economic output and what is needed for sustainability. So the bar for the success of Biosphere 2 was about the height of an ant.[7]

Additionally, the basic test was conceptually flawed because it ignored imports and exports. No life system that we know is sustainable without imports—in the case of Earth, imports of solar energy. However, for the moment, let's ignore the complexities added by trade and examine sustainability, focusing on the economic concepts of sustainability developed in earlier chapters.

AN OVERVIEW OF BIOSPHERE 2

Biosphere 2 was a huge glass structure, physically enclosed, near Tucson, Arizona, covering about 10,000 square meters (or about 2.5 acres). It contained many of the major Earth biomes, such as tropical forest, ocean, wetland, desert, and agricultural zones. It was stocked with a small number of biological species and sufficient resources to produce the food necessary to maintain eight humans for two years. It began with large stores of resources, drugs, and equipment costing around $200 million and imported a huge amount of energy (about $50,000 per person per year). For two years, the eight biospherians lived in this confined space, produced most of their food, and managed to survive.

From a technical point of view, the attempt to sustain a closed system was a failure. The major problem, life-threatening for humans, was a steady decline in atmospheric oxygen. The concentration went from 21% at the outset to a low of 14%, or just above levels that would be fatal. Depletion required a big infusion of oxygen, without which the humans were unlikely to survive. An important feature is that support staff were only a few feet away, ready to supply oxygen as needed. If that kind of catastrophic mistake took place on Mars, with a nine-month resupply time, no biospherians would have survived.

As perilous as the experiment was for humans, other species fared much worse. All pollinators (bees, etc.) went extinct. Of the 25 vertebrates at the outset, 19 went extinct. Most insects went extinct. There was one major successful survivor: the crazy ants ran wild. The crazy ant is a pest that has the distinctive property of being able to survive almost anywhere.

The eight people worked long hours, on average 10 hours a day, to maintain the viability of the system. Much of the time was spent on agricultural production—about 22 hours per week per person. By comparison, the average hours worked on farms in the U.S. economy equal about 0.1 hours per week. There was no reported production of any of the other major components of economic activity (shelter, clothing, transportation, drugs and health care, or entertainment). Economic output was therefore limited to sub-sistence agriculture.

MEASURING SUSTAINABILITY

How could we determine whether projects such as Biosphere 2 represent sustainable systems? The last two chapters had extensive discussions of sustainability. We need to adapt that discussion for this broader framework.

In thinking about sustainability, a minimum criterion is that the system is *economically viable* in the sense that it is productive—that outputs are larger than inputs. This is a straightforward concept and simply means positive net output. This is a low hurdle but useful as a starting point.

The preferred measure of sustainability is a system that is suffi-ciently productive to maintain its capital stocks. That is, the economy is sustainable if the stock of natural, tangible, and intellectual capital does not decline at current consumption levels.

The key concept here is *capital*. This concept denotes durable tangible or intangible goods that are used in production. Natu-ral capital includes forests and clean air; tangible capital includes equipment and houses; and intellectual capital includes patents, software, and technological knowledge. The total value of capital would be the quantities of each kind of capital times their prices or social values.

The sustainability of Biosphere 2 opens broader issues than the standard measures of sustainability. In our standard economic measures, we assume that certain parts of our natural capital are maintained. For example, we assume that the sun still shines, the rivers still flow, and most pollinators survive. We clearly cannot assume that for Mars or a planet light-years away. Including values for the fundamental parts of natural capital is beyond the scope of the present analysis, however, so we can take on the more limited task of looking at economic sustainability.

OUTPUT IN BIOSPHERE 2

To investigate the economic viability and sustainability of Biosphere 2, I have constructed a set of rudimentary economic accounts. These use the concept of national income accounting as discussed in the last chapter to measure net national product (NNP) and its components. The estimates are just suggestive, and perhaps others with access to better data can improve them. But here we go.[8]

The raw materials are the following: We have time-use data for the biospherians broken down by sector. The initial capital stock is estimated to be $200 million, with energy inputs at $0.8 million per year and security and other services estimated to be $0.5 million per year. Work hours for nonscientific activities are valued at $15 per hour in 2015 prices, while scientific activities are valued at $50 per hour. The major cost item is depreciation, which is estimated to be 10 % of capital per year, which is appropriate for equipment but probably low given that Biosphere 2 had a limited lifetime.

The one sector that might be highly valuable is investment in intellectual property or scientific knowledge. As is conventional, this is measured at cost, but there might be positive externalities that are much larger.

Table 10-3 shows the results and compares Biosphere 2 with the economic accounts of the United States in 2015 on a per capita basis. I emphasize that the estimates for Biosphere 2 are based on very fragmentary information, although the order of magnitude is surely correct. The first five lines show gross output by industry—that is, simply

TABLE 10-3. Estimated economic accounts for Biosphere 2 and the United States

Sector	Per capita output (2015$)	
	United States	Biosphere 2
Farms	1,256	1,005
Industry	2,615	0
Trade	0	0
Services	33,607	23,166
Investment, intellectual property (scientific knowledge)	952	18,876
Government and other	1,182	0
1. Per capita gross output	**98,083**	**43,047**
Less: intermediate inputs	41,998	233,142
2. Per capita gross domestic product	**56,084**	**−190,095**
Less: Capital consumption and other	8,178	3,252,969
3. Per capita net national product	**47,907**	**−3,443,064**

Data for US are for 2015.

Note: Estimates show the per capita output of the two economies. Estimates for Biosphere 2 are for the period 1991–1993 while the United States is for 2015. All estimates are in 2015 prices and wages.

what is produced, such as the value of carrots. The estimated value of per capita gross output in Biosphere 2, shown in line 1, was about one-half of that of the United States. Note that the output was highly imbalanced, with zero production of industry or trade. The most important outputs were services and creation of intellectual property.

Line 2 shows the GDP of Biosphere 2, which equals gross output less inputs such as energy. My estimate is that inputs exceeded output (even energy inputs were greater than outputs), so the GDP of Biosphere 2 is estimated to be minus $190,000 as compared to plus $56,000 for the United States.

The final total shown in line 3 is NNP in Biosphere 2, which equals gross national product less the depreciation of capital. Our earlier discussion identifies NNP as sustainable income. Depreciation of Biosphere capital is estimated to total more than $3 million per person per year. Subtracting depreciation gives a final calculation of per capita NNP of minus $3.4 million per year. We can run variants of these numbers, but they always add up to very large negative numbers.

The Verdict on the Sustainability of Artificial Earths

What do we conclude about the prospects of establishing artificial Earths or sustainable human systems in remote places? Our discussion of the history of human civilizations recounted the long road to today's highly productive global economy, which suggests formidable barriers to establishing a parallel system in a remote and hostile location. Moreover, in reviewing the prospects of colonizing Mars or other planets, the conclusion was similarly pessimistic. Based on analogous situations, such as Antarctica or the ISS, the costs of maintaining life in hostile circumstances look astronomical.

The results are even more pessimistic when we look at the history of Biosphere 2. It failed miserably the test of economic viability and the test of economic sustainability. Even when situated on Earth, the system could not sustain a modern living standard, or even a Paleolithic living standard. If operated for long, it would run down to zero. Everything and everyone inside the biospheric bubble, except perhaps the crazy ants, was doomed.

The lessons from these three experiments are consistent and humbling. For the near future, the prospects of a self-sustaining exo-civilization are remote.

11

Pandemics and Other Societal Catastrophes

Observing societies at the beginning of 2021, we see people around the world getting sick from COVID-19, being afraid of getting sick, and even dying by the tens of thousands every day. They shelter as best they can. They look for hope as two new highly effective vaccines have been approved, but they worry because supplies are limited and distribution is hampered. In short, the world encountered a societal catastrophe when the "novel coronavirus" emerged from a live-animal market in Wuhan, China.

As a matter of terminology, the novel coronavirus (also called by its scientific name SARS-CoV-2) is the deadly virus that started circulating the globe in January 2020. The virus causes a complex set of diseases generally called COVID-19. The two are often used interchangeably, but for simplicity I will refer to the pandemic as COVID-19 except when necessary to distinguish the disease from the virus.

A *societal catastrophe* is an event that causes widespread social, economic, and political hardship. In the modern era, such catastrophes are rare, occurring with a frequency of decades, centuries, or even longer. Because they are *low-probability, high-consequence events*, they pose particular difficulty for social decision processes—for detection, prevention, and mitigation. Indeed, as we see in the case of the COVID-19 pandemic, our attitudes and political institutions make it extremely difficult to deal effectively with catastrophes, even when we have the most advanced technologies available.

The 2020 pandemic is a catastrophic viral plague. But at other times, we worry about catastrophes such as nuclear conflict, earthquakes, asteroids, and climate change. Most of us normally spend little of our time worrying about catastrophes—but when they occur, we can think of little else.[1]

Categories of Catastrophes

Catastrophes have different levels of severity, extent, and speed. At the less severe end are regional wars and famines. More destructive are world wars or severe pandemics that cause widespread death and destruction. The worst nightmares, such as from large asteroids or potential nuclear war, would lead to devastation of large parts of Earth and degradation of human existence.

Catastrophes may be regional (like limited wars) or global (like pandemics and climate change). Some hit us very quickly, as was the case with the COVID-19 pandemic, while others, like climate change, build gradually over years or decades. A final dimension is their frequency. Some catastrophes, such as regional wars or minor epidemics, occur every few years. Others, such as the dinosaur-killing asteroids, are much rarer and occur at a frequency of tens or hundreds of millions of years. The challenge of rare events will be considered again shortly.

The Green Dimension of Catastrophes

Environmental science and economics study spillovers or externalities: pollution, climate change, nuclear fallout, dead fish, and dying oceans. Catastrophes like pandemics fit into a book about externalities and Green policies because they are a particularly frightening externality. Epidemiologists calculate that for COVID-19, in a world without protections or mitigation, each infected person infects about three other people. Perhaps 10% of those people become gravely ill and 1% die. If we are infected and cough or shout or sing, we are polluting the air around us with a deadly virus. Taking precautions protects ourselves and families, but it also protects our friends, and even strangers.

Governments have a central role in dealing with deadly externalities such as infectious diseases. Agencies such as the U.S. Centers for Disease Control and Prevention (CDC) and its equivalent in other countries have developed detailed protocols for dealing with infectious diseases—not just endemic ones such as the flu but also new ones such as COVID-19. These agencies are like the U.S. Environmental Protection Agency, with specialized scientists and a tool kit to deal with public-health crises.

But the current crisis shows that scientific expertise cannot by itself stop a pandemic. Political leaders have a central role to play in shaping public opinion and adopting appropriate policies. In this pandemic, the leaders of China and the United States failed at their jobs. China's leaders hid the outbreak and were slow to alert their own people and the rest of the world. President Trump displayed willful ignorance and political self-interest, thereby hampering the U.S. response. We will never know how many people needlessly died because of the failure of political leadership, but the COVID-19 crisis illustrates the need for Green attitudes and policies for the most severe of external impacts, as well as more routine ones.

Putting this differently, humans are vulnerable but not helpless in the face of catastrophes. In each case we can take preventive or precautionary steps to avoid and mitigate the catastrophes. The case of catastrophic climate change is a clear example here. If earth scientists find that disaster lies beyond a certain temperature trajectory, nations can take steps to avoid crossing that threshold. A combination of strong emissions reductions policies, such as high carbon taxes, plus energetic investments in renewable technologies can bend the curve.

Different approaches are needed for different catastrophes, but the widespread mayhem that has allowed the spread of COVID-19 was not inevitable. Our fates lie, not in the stars, but with ourselves and our political leaders and institutions.

Extreme Events

Some catastrophic events, like hurricanes in Florida, are terrifying but not surprising. Other events are extremely unlikely—indeed, so unlikely that most people ignore them. They might be unlikely because they are simply rare, like huge asteroids landing on Earth. Or they might be unlikely because they are new to Earth or human experience—one example being the detonation of the first atomic weapon over Hiroshima in August 1945. Nothing in human experience could have predicted the vast destruction. The COVID-19 pandemic of 2020 was unexpected because this particular virus had never infected people before nor had its genetic code sequenced.

These extreme events are sometimes called *tail events*. A tail event is an outcome that, from the perspective of the frequency of historical events or perhaps from intuition, should happen only once in a million or billion or centillion years. It is called a tail event because probability distributions have tails (think of the two ends of a bell curve), and a tail event is one that is extremely far out the tail.

Consequential or catastrophic tail events, such as the emergence of the COVID-19 pandemic, are particularly difficult to deal with primarily because they are so unpredictable. The result is that we fail to invest in the programs to prevent them, to slow their spread, or to mitigate their damage. Some of the worst societal catastrophes fall into the category of tail events.

Dealing with tail events does not change the basic requirements for mitigating and preventing societal catastrophes. It does add another complexity—that of dealing with low-probability events. But the basic requirement of sound science, political leadership, and institutions lies at the heart of cutting the tails off the worst catastrophes.

The Challenge of Pandemics

Catastrophes are woven throughout human experience from the beginning of recorded history and before in fables and myths. Table 11-1 shows the most lethal pandemics for which we have

TABLE 11-1. Plagues old and new

Rank	Name	Year start	Year end	Fatalities (000)	Fatalities as share of global population
1	Black Death	1331	1353	137,500	38.261%
2	Plague of Justinian	541	542	62,500	32.094%
3	Antonine plague	165	180	7,500	4.048%
4	Spanish flu	1918	1920	58,500	2.768%
5	Cocoliztli epidemic of 1545–1548	1545	1548	10,000	2.367%
6	Third plague pandemic	1855	1960	18,500	1.600%
7	Smallpox epidemic in Mexico	1520	1520	6,500	1.538%
8	Japanese smallpox epidemic	735	737	2,000	0.967%
9	HIV/AIDS pandemic	1920	2020	30,000	0.882%
10	Plague of Cyprian	250	266	1,000	0.532%
11	Cocoliztli epidemic of 1576	1576	1580	2,250	0.444%
12	Plague in Kingdom of Naples	1656	1658	1,250	0.226%
13	Persian plague	1772	1772	2,000	0.221%
14	Plague of Athens	−429	−426	88	0.191%
30	COVID-19	2019	?	1,750	0.022%

The table shows a compilation of the most lethal plagues on record. The last column shows the fatalities as a proportion of the global population. Note that by early 2021, the COVID-19 pandemic had already reached number 30 on the list of most deadly and was nowhere near burned out.[2]

reasonably reliable estimates (although those are only approximate before the last century). The last column shows the deaths as a proportion of the estimated world population. Early pandemics, such as the Black Death of the fourteenth century, would wipe out virtually the entire population of a region. In the last century, the "Spanish flu" of 1918–1920 and HIV were major killers of the population. The row for COVID-19 is a question mark as this book is finished in early 2021.

One preliminary conclusion from table 11-1 is that modern science and medicine have prevented the worst health catastrophes of the prescientific age, but they have so far not removed the periodic and unpredictable emergence of lethal contagious diseases.

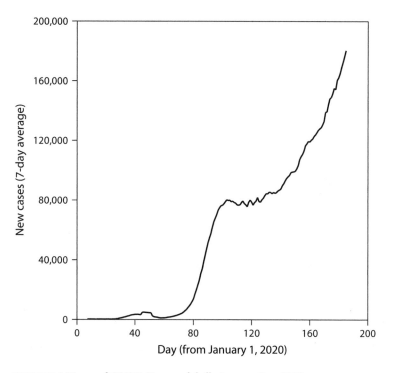

FIGURE 11-1. Measured COVID-19 cases globally, January–June 2020

The COVID-19 Pandemic

It will be useful to provide a brief overview of the current pandemic before putting it in a Green perspective. The novel coronavirus emerged in late 2019 and was most closely linked to a virus found in bats. Patient zero (the first documented case) dates from December 2019 in Wuhan, China. Chinese medical authorities became aware that a new virus had emerged in early January, and the sequenced gene was posted on January 11, 2020.

The virus spread rapidly around the world over the next three months (until day 90, or the end of March), as seen in figure 11-1. The first surge took place in January 2020, with a decline in February. During the early stages, the number of cases was doubling every three or four days. There was a decline in global cases as China locked down its population. The next surge took place starting in mid-March, exploding in the United States and Western Europe. A second pause occurred as affected countries locked down their businesses and households (between day 70 and 90). Then, as countries started opening up, cases started growing again in early May and continued to grow rapidly until this book was completed in early 2021.

Infectiousness and Lethality

To understand pandemics requires explaining the most important characteristics of deadly agents: their infectiousness and their lethality. By *infectiousness*, we mean the number of people on average that an infected person will infect if there are no measures to prevent the spread of the disease. This goes by the term $R0$. The R is the reproduction rate between "generations" of infected people. The 0 in $R0$ refers to the infectiousness of patient zero or of those infected before slowing measures are taken. For example, the $R0$ for COVID-19 has been estimated to be around 3, although some variants may have higher $R0$. Suppose that $R0 = 2$. If 1,000 people have an infection in generation n, then 2,000 will have it in generation $(n + 1)$, again assuming no protective measures are taken.

The other important feature of an infection is its lethality, which we will call L. This represents that fraction of people who die after contracting the illness. At the low end of lethality is the common cold, with a death rate of close to zero. At the upper end is smallpox, with L of about 30%. The lethality of COVID-19 is still uncertain, but estimates range between .5% and 2% of infections.

The infectiousness and lethality of COVID-19 can be compared with that of the "Spanish flu" of 1918–1920. Recall from table 11-1 that this illness killed about 3% of the world's population. In today's world, that would be more than 200 million people.

Figure 11-2 shows the combination of lethality and the infectiousness of several diseases. It is important to note that infectiousness numbers (measured by $R0$) are for *unprotected and unvaccinated* exposures. Smallpox is one of the worst diseases if unprotected because it is both highly infectious and lethal; it wiped out vast numbers of indigenous Americans when Europeans arrived in the 15th century (see table 11-1). Today, there are literally zero fatalities

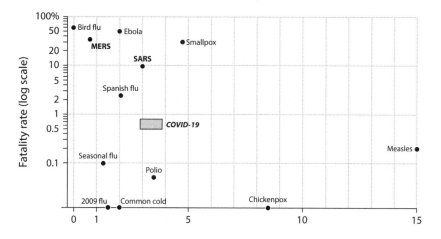

FIGURE 11-2. Infectiousness and lethality of COVID-19 and other diseases in unprotected populations
Source: From Knvul Sheikh, Derek Watkins, Jin Wu, and Mika Gröndahl, "How Bad Will the Coronavirus Outbreak Get? Here Are 6 Key Factors," *New York Times*, February 28, 2020. Updated based on Nicholas Christakis, personal communication.

from smallpox because of the combination of an effective vaccine and a long-term public health campaign to wipe it out.

To understand the control of pandemics, we need one further concept, the *effective reproduction rate*, which I will call *Reff*. Recall that *R0* is the rate of reproduction in a completely vulnerable and unprotected population—for example, before the virus has been detected. However, once protective measures are taken, the reproduction rate falls. It might be reduced because infected individuals are isolated, because people shelter in their houses and are not exposed to infected individuals, or because people are immune due to prior infection or vaccination.

When effective protective measures are in place, *Reff* will be below *R0*. The key to combating any pandemic is to reduce *Reff* below 1. Suppose, for example, that 1,000 people have been infected, and *Reff* has been reduced to .5. Then the number of infected people will decline by 50% each generation. The pandemic might even disappear if there is no reservoir of viruses.

If we measure *Reff* using COVID-19 case data, the number was extremely high during the rapid growth phase in January through

March 2020. Then, as cases stabilized in the summer of 2020, *Reff* was close to 1. However, as cases grew rapidly at the end of 2020, the *Reff* climbed again and cases were doubling every 2¼ months. Public health specialists hope and expect that when a large fraction of the population is vaccinated, sufficient immunity will be reached (called "population immunity") that the pandemic phase will be over.

Reducing Infectiousness and Lethality

Why are there so few cases of smallpox, Spanish flu, measles, and polio today? The reason is that health measures have reduced both the lethality and the infectiousness of these diseases. Smallpox is easy to understand. There are no cases of smallpox today. Even if *Reff* is high, the number of cases would be *Reff* times zero, which is always zero. In the case of measles, the vaccine is highly effective, so the actual infectiousness is close to zero.

In the case of COVID-19, the *Reff* can be reduced *temporarily* by social distancing—that is, reducing the number of contacts with potentially infected people. That will slow the disease as long as we stay distanced, but *Reff* will pop back up when we mingle in bars and stadiums. The way to reduce *Reff* permanently is with an effective vaccine, which reduces infectiousness because there are fewer people for infected people to infect.

There is one final way to reduce *Reff* and that is to achieve "population immunity" (sometimes called *"herd immunity")*. This occurs when a sufficient number of people who have had the disease develop immunity. Suppose that *R0* is 2 in a uniform population, and further suppose that three-quarters of people are immune because they are vaccinated or have antibodies from a prior infection. Then *R0* becomes $\frac{1}{4} \times R0 = \frac{1}{2}$, which means the infections will die out as in the example above.

Some ignorant political figures have advocated allowing cases to continue until the world achieves herd immunity by infection rather than vaccination. This is a gruesome scenario for COVID-19 because it would require more than five billion people to get the illness to achieve herd immunity on a global scale.

Dealing with Pandemics

Pandemics pose great difficulties because they strike suddenly and unexpectedly and move so swiftly. They challenge our institutions to prepare in advance and to execute plans more swiftly than the infections spread. In the case of COVID-19, the disease was clearly spreading in the population before it was identified.

The situation in New York City will illustrate the point. The first case in NYC was announced on March 1, 2020. By April 1, the number of cases had grown to 54,000. Later testing of blood samples for antibodies to COVID-19 found that the number of infections in this period was 642,000. This larger number represents the actual number of cases in the NYC population. This suggests that there were hundreds of infections in the population before the first case was reported.[3]

The COVID-19 pandemic had four features that made it particularly dangerous. The first two were its high lethality and infectiousness, as shown in figure 11-2. But two others were also critical. One was the fact that it was susceptible to human-to-human transmission. But this critical fact was only discovered in the last half of January.

A final critical factor is that COVID-19 allows both asymptomatic and presymptomatic spread. Other diseases with asymptomatic infections are typhoid, HIV, and cholera. But many diseases are transmitted only when the infected person shows symptoms. The evidence of antibody testing as well as contact tracing suggests that a large number of infected people are asymptomatic. Some of these—it is not known how many—can transmit to others without warning. One major issue is that the asymptomatic transmission of COVID-19 provoked confusion even as late as June 2020.[4]

For many Green issues, such as global warming, humans have had many years to study and prepare responses. Pandemics do not allow a relaxed and reflective reaction. When cases are growing at 200 % or 500% per week, taking time to ponder the best response will allow a virus to overwhelm the globe.

In fact, different levels of government in the United States have planned for pandemics in recent decades. They have generally focused on influenza viruses since those have been the major threat over the last century—indeed, all major pandemics of the last century have been viruses.

The CDC, which is the lead U.S. agency that deals with pandemics, has issued a series of manuals with steps for preparedness at governmental and private levels. Additionally, the CDC has worked to coordinate with the World Health Organization, which also has strong capability to monitor infectious diseases. China has built up its own CDC. Each of these agencies has issued guidelines, procedures, and reports with useful guidance. However, in the face of the storm, neither the Chinese nor the U.S. CDC followed their guidelines and procedures. A week or two of delay in a pandemic is lethal for a country and even for the globe.

Criteria for Assessing Pandemics and Other Catastrophes

When catastrophes occur, scientists and historians need to look back and assess the successes and failures of policies. Four key attitudes and policies are necessary to deal with catastrophes like pandemics:

- Adequacy of relevant scientific and technological expertise
- Level of preparedness
- Effective execution
- Effective communication by leaders in the public and private sector

As of fall 2020, the United States had the largest number of cases and deaths of any country in the world. Many people are wondering how the wealthiest country on Earth, which leads the world in science and technology, could fail so miserably to respond effectively to the COVID-19 crisis.

It is too early to provide a full assessment of policies at the time of this writing since we do not know the state of the pandemic in a year, or two years, or five years, or more—nor do we know whether it will linger and flare up in the decades to come. A complete appraisal must await the course of the pandemic. We can, however, gather the information at hand and make a preliminary judgment as of the fall of 2020.

Science and Technology

The United States has great resources in science and technology. In comparison to the state of knowledge a century ago, during the Spanish flu, we have incomparably better understanding of the underlying disease in 2020. As an example, for many years it was thought that influenza was bacterial rather than viral. Indeed, it was not until 1944 that influenza was categorized as a virus. By contrast, the current coronavirus was sequenced and published for the world on January 11, 2020—less than two weeks after Chinese doctors identified it as a new viral strain.

One of the stunning scientific developments of 2020 was the ability to develop, test, and gain approval for two vaccines to prevent COVID-19 illness. Others are in the pipelines, but at the beginning of 2021, people had received tens of millions of doses of a major new technology for vaccine development, with the prospect of global vaccination of the willing within two or three years. If the pace of vaccination occurs as hoped and the benefits are as anticipated, the United States and other major countries are likely to achieve population immunity in a year if all goes well. The long nightmare of the plague will not be over, but the worst will be in the rear-view mirror. Then and only then can the global economy and society return to their normal condition and people once again feel secure in travel, mingling, and social closeness.

Preparation and Resources

Public health experts in the United States and other countries—as well as the World Health Organization (WHO)—have long known of the dangers of pandemics. They have frequently prepared plans to deal with them. These plans contain most of the elements that evolved in the United States and other countries over the course of the

TABLE 11-2. Fiscal resources to cope with different threats, United States, 2021

Category	Funding, 2021 [mllions of $]
Department of Defense	740,500
Federal Research and Development	142,200
Health Research	36,965
Centers for Disease Control	12,612
Influenza Planning	40

Source: Budget of the US Government, Fiscal Year 2020.

COVID-19 pandemic, such as testing, contact tracing, social distancing, and border controls, as well as active and passive surveillance.

However, the best of plans will do little without the resources to provide the necessary staff and programs to implement crucial public health measures. Perhaps the clearest evidence of lack of preparation in the United States was the amount of resources devoted to pandemic planning. For many years, conservatives have pushed to "starve the beast," where the beast is nondefense governmental spending. The "beast" includes pandemic planning. This is seen in the federal budget. One major category is "Countering Emerging Threats," which lists several topics but not pandemics. The only mention of pandemics is buried in a paragraph with three words, "International Pandemic Preparedness."[5]

Table 11-2 provides an overview of the funds devoted to different threats in the federal budget in 2021. The military budget was $741 billion. The lead agency for tracking pandemics is the CDC, and its budget was $12.6 billion. However, of that, only $40 million was for pandemic planning and that was targeted on influenza. This sum amounts to about ten cents per year per person. The United States spends one thousand times more on pet food than on pandemic preparation.

The picture here reminds us of the words of the economic historian Joseph Schumpeter:

The spirit of a people, its cultural level, its social structure, the deeds its policy may prepare—all this and more is written in its fiscal history, stripped of all phrases. He who knows how to listen to its message here discerns the thunder of world history more clearly than anywhere else.[6]

And the thunder we hear is that American fiscal policy completely ignored the major threat to its society coming from new diseases.

Implementation

One of the American disadvantages in fighting pandemics is its federal structure. Many of the most important decisions and legal authority on public health reside at the state and local levels while the federal government has resources, expertise, and central command.[7] The CDC organizes and authorizes testing but has tiny fiscal resources; states and localities have authority for shutdowns and quarantines but are uncoordinated, have little expertise, and are perennially fiscally stressed.

The pandemic response in this respect differs 180 degrees from a military one. If the enemy is aircraft or troops, the federal government has massive funding, a large army, and clear lines of command; if the danger is from a tiny virus, the authority is poorly funded, understaffed, and widely dispersed.

The importance of implementation can be seen in the failed rollout of COVID-19 testing. The CDC had the facilities and expertise to manage testing, but it botched the effort from the start. There were four major failures. First, its initial tests were defective, and it took weeks to correct the failure. As a result, the United States was weeks behind other countries. Second, the CDC refused to allow other entities, such as hospitals, to devise their own tests. Many private entities had the capabilities, but the scientifically conservative approach of the CDC centralized testing in its own hands, and its hands were shaky. Third, the CDC did not allow widespread population testing to determine the overall prevalence of COVID-19. Later studies found, for example, that there were thousands of cases in the New York City area by the time the first cases were confirmed. A final mistake, again the result of scientific conservatism and overcentralization in the CDC, was to prohibit pool testing, in which tests could combine the samples of multiple individuals. This technique, which is most valuable when the prevalence is small, would have been particularly useful at the early stages when there were few confirmed cases, and tests were extremely limited.

This single example of a bungled testing strategy was devastating for the United States. It allowed the virus to grow rapidly and virtually undetected from February to April. By the time testing began to catch up, there was widespread community spread across virtually the entire country. Even though testing grew rapidly, it could not keep up with infections.

Communication

Without doubt, the major failure of the United States was in communication. Here, the pandemic strategy of the G. H. W. Bush administration put the point clearly:

> A critical element of pandemic planning is ensuring that people and entities not accustomed to responding to health crises understand the actions and priorities required to prepare for and respond to a pandemic. [This requires] clear, effective, and coordinated risk communication, domestically and internationally, before and during a pandemic. This includes identifying credible spokespersons at all levels of government to effectively coordinate and communicate helpful, informative messages in a timely manner.[8]

In reality, the response of the U.S. federal government under Donald Trump was a case study in mismanagement, with mixed messages, denial of reality, false statements, and ludicrous predictions from the president and those politically infected by him:

> On February 27, the President stated, "It's going to disappear. One day it's like a miracle—it will disappear." Within two months, the number of cases had multiplied 100-fold.
>
> On March 6, the President stated, "Anybody that needs a test, gets a test." On that day about 1700 people were tested in a country of 330,000,000 people.

On March 26, the President stated, "Nobody would have ever thought a thing like this [a pandemic] could have happened." In fact, public-health experts had repeatedly warned of pandemics and prepared reports on response. For example, a 2019 government report said that "the United States and the world will remain vulnerable to the next flu pandemic or large-scale outbreak of a contagious disease."

On April 24, the President stated, "I see the disinfectant, where it knocks it out in a minute. One minute. And is there a way we can do something like that, by injection inside or almost a cleaning?" (Some of the President's recommended disinfectants, if injected or swallowed, are lethal.)

On March 23, the President stated, "But we've never closed down the country for the flu. . . . So you say to yourself, 'What is this all about?'"

The point is not to highlight President Trump's uneasy relationship with the truth. We know from interviews with Bob Woodward that he was aware of the dangers and lethality of the coronavirus in early February. Trump's concerns about his political future and reelection in the 2020 election completely outweighed his responsibilities as president.[9] The result was that federal leadership in the United States did exactly the opposite of what was necessary. The communication was confusing instead of clear, muddled instead of effective, and disordered instead of coordinated.

In addition, public health became politicized as a strategy to deny accountability. Trump politicized the virus from the outset, with the resulting politicization of science and policy. The most damaging aspect was to politicize the wearing of masks and turn a politically neutral and highly effective public-health measure into a political flash point.

The intrusion of *politics* into public health is dangerous. It prevents forming a national consensus on taking critical steps. The growth of cases might be slowed sufficiently by universally using face masks, avoiding crowded spaces, and closing dangerous venues such as large athletic stadiums, bars, and casinos. Moreover, taking such steps might allow the rest of society and the economy to function in near-normal fashion. But if a substantial fraction of the population believes that the virus is a hoax, that face masks do not work, and that their civil liberties include going to bars and large gatherings and not wearing face masks, then the steps necessary to slow the spread of the disease may be much more intrusive and costly.

The light at the end of a dark tunnel

This concludes our introduction to social catastrophes, focusing on the 2020 COVID-19 pandemic as a case study. It seems likely that the world will soon turn the corner from the social, economic, political, and health nightmares of 2020 to the beckoning but distant light of a normal life. If Fortune smiles on humanity, we will begin to enjoy our daily lives—mingling and working and going to school and taking vacations—in a year or so. With effective vaccines and a successful public-health campaign, such as was accomplished for smallpox and measles, the new coronavirus will gradually recede into the health background as something like the flu.

But the painful lesson of 2020 must be remembered. Societal catastrophes will come again in one form or another. We must prepare for them rather than ignoring them and suffering from their worst consequences.

PART III

Behavioralism and Green Politics

Behavioralism as the Enemy of the Green

"The fault lies not in our stars, but in ourselves," was the way Shakespeare's Julius Caesar characterized his political troubles. Our environmental troubles, similarly, sometimes lie not in misbehaving markets but in people's flawed decisions. These are usually called *behavioral anomalies*, which is a fancy term for a class of harmful private activities that appear to be lazy, uninformed, or perverse.

The interesting point about behavioral anomalies is that inefficiencies arise from private actions rather than market failures. Suppose you are playing basketball, and you systematically miss easy shots. You cannot blame it on your school or league. Perhaps you are not paying attention. Or your technique is poor, and you will not listen to your coach. Whatever the reasons, you have a low score.

Similarly, psychologists and economists have identified low scores with respect to a wide range of personal decisions. One of the best-documented anomalies is the excessive use of energy. Also, people do not pay sufficient attention to price signals (and therefore may not respond to environmental policies). Irrational addictions to alcohol, drugs, texting, and speed can produce carnage on the highways. Sometimes, people just appear to act haphazardly, and this can lead to harm to themselves or others. The point is that behavioral anomalies can either produce or worsen harmful side effects, and in some cases these may be lethal. So we need to add poor decisions to the list of issues that Green policies address.

Behavioral anomalies have a Brown tint in two important cases. The first are biases that lead to excessive energy consumption and pollution. Such biases are endemic in our economy and are illustrated by what economists call first-cost bias, examined later in this chapter. A second case is waste that arises from inefficiencies. These are not necessarily biased, as in the first case, but lead to excessive pollution because they use more resources than are necessary (perhaps overusing first-growth forests and clean water along with excessive labor and capital).

Before dealing with examples of decision failures, let us review the background. Economists and psychologists have long puzzled over the reasons for behavioral failures. It should be emphasized that decision failures are not limited to Green sectors. People make mistakes in many areas. They sometimes make poor decisions about their health (they do not take their medications), about finances (they do not read the mortgage document and lose their homes), and about business (half of small businesses fail in their first year). Such anomalies have been studied fruitfully by psychologists Amos Tversky and Daniel Kahneman, economists George Akerlof, Robert Shiller, and Richard Thaler, and lawyers such as Dan Kahan and Cass Sunstein.

Looking for behavioral anomalies and flawed decision-making has become a major topic in psychology and economics (the latter being the field of behavioral economics). Scientists have identified more than a hundred different anomalies, from action bias to the verbatim effect. Every time something strange happens, it is seen as an example of behavioral economics at work. Two important examples are defective discounting and first-cost bias.

Anomalies of Discounting

One of the central issues discussed by behavioral economics is the role of *discounting*. Standard economics holds that people should use market rates of return to value investments, such as those on purchasing automobiles or energy-efficient houses and appliances. A substantial body of research indicates that people use too-high discount rates and that investments are biased toward too much energy use.

Here is the issue in a nutshell: Many actions require investments today to reduce costs in the future. For example, when we make investments to reduce pollution, these costs are paid largely in the near term. However, the benefits in the form of reduced damages may come far into the future. Suppose that we replace a coal-fired power plant with a wind farm. If we follow the chain of effects from building the wind farm to reduced sulfur emissions to reduced damages, there is a delay of many years or decades from building the wind farm to the reduction in damages.

Discounting is important because it allows us to put future and present dollars in the same units. Suppose that I make roof repairs that cost $1,000 today, with the result that I save $2,000 on a new roof in 10 years. Is this a good investment? To answer this question, we need to put all the dollars on a common footing. This is done by translating all dollars into a *present value*. This tells you the value in today's dollars of the stream of inflows and outflows. We take the future dollars and convert them into today's dollars using a discount rate.

For example, suppose that the average rate of return on investments is 5% per year. I might therefore use a discount rate of 5%. Investing $1228 for 10 years at 5% per year gives $1,228(1.05)^{10} = $2,000$. Similarly, $2,000 in 10 years is worth $2,000/(1.05)^{10} = $1,228$ today in present value.

So, going back to the roof example, adding up all the costs (negative) and benefits (positive) yields a present value of $-$1,000 + $1,228 = 228. So, at a discount rate of 5%, my roof repairs would be a sound investment.

However, suppose that I use a much higher discount rate—in effect, I think money in the future is worth less than it actually is, or I "overdiscount" the future. I might use a 20% annual discount rate in the calculations. At this rate, $2,000/(1.20)^{10} = 323. If we sum the discounted flows, they are equal to $1,000 + $323 = -$677$. So, at a discount rate of 20%, the present value of the investment is actually negative. Applying the superhigh discount rate, the investment does not pay off using standard financial analysis.

This hypothetical example appears widespread in household decisions about energy and other investments. Here are some examples as summarized by George Loewenstein and Richard Thaler:[1]

> A study comparing pairs of refrigerators differing only in energy use and initial purchase price revealed that the implicit discount rates associated with purchasing the cheaper models were incredibly high: from 45 to 300 percent. [Another study] computed the discount rates implicit in several different kinds of appliances. They found that the implicit discount rate for room air conditioners was 17 percent. However, the discount rates for other appliances were much higher, e.g., gas water heater, 102 percent; electric water heater, 243 percent; and freezer, 138 percent. Economic theory has a clear prediction about these inefficient appliances—they will not be produced. But they are produced, and purchased.

Another common anomaly is *hyperbolic discounting.* This finds that people apply discount rates that are much higher in the near term than in the longer run, so the discount rate looks like a hyperbola rather than a constant. Here is how a pioneer in the field, David Laibson, explains it: "Hyperbolic discount functions imply discount rates that decline as the discounted event is moved further away in time. Events in the near future are discounted at a higher implicit discount rate than events in the distant future."[2]

Hyperbolic discounting can be seen as an example of the "now, future" dichotomy. We want our pleasures now and care less about the future, but we do not distinguish the near future from the far future. The main effect here is to overdiscount the future—to place too much weight on present costs and too little weight on future benefits. Decisions subject to hyperbolic discounting (or overdiscounting) systematically undervalue the future, and this means too little Green investment.

The Syndrome of First-Cost Bias

A second major behavioral issue, particularly hazardous for long-lived investment, is myopic choices among alternative designs. This is known as first-cost bias in economics. First-cost bias has been documented again and again. It leads to investments that use too much energy and therefore lead to excessive environmental impacts associated with energy use.[3]

Here is a little fable that relates to housing but could equally well apply to air conditioners or cars. Suppose I am considering how to insulate my home. The builder shows me two approaches. One is a standard fiberglass roll that is easy to install; the other is a rigid foam board, which is more expensive but has almost double the insulation value. The fiberglass batts cost $5,000 to install, whereas the foam board costs $7,000. The upfront installation costs are easy to calculate.

However, calculating the savings is more difficult. To determine which investment is better, I need to know the energy use for each kind of insulation. My contractor tells me that foam insulates almost twice as well as fiberglass and provides me the technical numbers on each. It would take some calculations to determine what the savings are in my climate zone.

I turn to an expert engineer who runs the numbers and estimates that the energy costs are $500 per year for the high-investment foam and $900 per year for the low-investment fiberglass.

At this point, I might give up. It is all too complicated, and I do not have the time or skill to make the calculations necessary to determine what is best. Or I might suffer from the bias of hyperbolic discounting. People also may have trouble keeping on top of their credit-card debt, or face steep college tuition bills for their children, or have high medical bills. For any of these reasons, the extra $2,000 for the extra insulation is not welcome. They will end up minimizing the first cost. This decision leads to higher energy bills along with the resulting pollution.

Sources of Behavioral Anomalies

Why do people systematically make poor decisions, such as those seen in overdiscounting or first-cost bias? This question has been extensively studied, and there is no single answer. Here are some of the prominent reasons.

INFORMATIONAL PROBLEMS

People sometimes have incomplete information or do not process information efficiently. In our insulation example, it turns out to be exceedingly difficult to get sufficient information to make a good decision on the optimal insulation in a house. You can try to calculate the present value of the two investments in your head, but that is beyond most people's capabilities. In many cases, the information is simply not available. For example, when I am considering replacing my current refrigerator, I do not know how much electricity it uses. If people ignore what they do not know, it can lead to ignoring the difference in future costs.

DECISION PROBLEMS

While classical economics assumes that people act rationally, we know that people make all kinds of trivial and tragic decision mistakes in daily life. One example, called the credit-card anomaly, comes in people's financial decisions. Many people have savings in the bank with an interest rate of 1% per year, yet they borrow on their credit card *month after month* at an interest rate of at least 19.99% per year. Consumers are paying tens of billions of dollars of unnecessary interest. Are they unaware of the interest rate differences? Are the interest costs too small to notice in the credit-card statement? Perhaps there is an explanation analogous to first-cost bias in the insulation example, in which people might think, "Look, I know that I am paying $300 a year in interest, but that's less than a dollar a day." Whatever the reasons, these decision failures cost people boatloads of dollars.

INSTITUTIONAL PROBLEMS

Often, institutions shield people from price incentives and prevent them from making sound social decisions. An example is the absence of energy metering, such as at colleges. For almost everyone involved—students, faculty, and staff—the people who pay the bills are not the people who make the decisions. Students are inclined to load their rooms up with electricity-using devices (music systems, TVs, computers, microwave ovens, refrigerators, and so on). But student dormitories are almost never metered, so students treat energy as a free good. Scientists in laboratories almost never get charged for their energy use, so they have no incentive to buy energy-efficient equipment. This was discussed earlier under the rubric of principal-agent problems, where inefficiencies arise because the people who make decisions are different from those affected by them.

NONECONOMIC PREFERENCES

Economists often assume that people choose and use goods and services to maximize their own personal self-interest. This assumption is behind the invisible-hand principle about the efficiency of competitive markets. Laboratory experiments, market research, and common sense tell us that this assumption is often inaccurate and that people have noneconomic or even weird preferences. One example of strange preferences is hyperbolic discounting, which leads people to underweight future events sharply. Under hyperbolic discounting, there is today and the future, and the future is worth very little whether it is the next month or the next decade.

Researchers have identified many nonstandard preferences, such as status quo bias, loss aversion, uncontrollable passion, and rage, along with altruistic, erratic, and random behavior. People sometimes buy things "to keep up with the Joneses," not because they actually want them. People used to covet huge gas guzzlers with long tail fins and lots of chrome. Today, people might buy a tiny electric car because they want to help the environment.

These four categories of reasons for behavioral anomalies are important features of our economic and environmental lives. Even if markets are perfect, without monopolies or externalities, poor decisions can cause personal and environmental harms.

Solutions to Behavioral Anomalies

Behavioral anomalies pose thornier problems than externalities. This is because the sources of behavioral anomalies differ from and require different remedies from pollution-type situations. Suppose that first-cost bias in selecting cars arises because of informational deficiencies; the remedy would be to provide better information. On the other hand, suppose that the problem is machoism, and people want something resembling a tank. We might try to dissuade them with high gasoline taxes, or taxes on gas guzzlers, but information will not change their minds. For students who do not pay the energy bills, colleges might prohibit energy-intensive devices, but such bans are unpopular and hard to enforce. More effective than unenforced regulations would be to put electricity meters on student rooms where they pay for excess use.

Sometimes, people just need "nudges" to overcome inertia or because they know so little about the subject. The theory and practice of nudges has been an important addition to modern policies toward behavioral anomalies.[4]

These examples suggest that the remedies for behavioral anomalies tend to be specific to the anomaly and the sector. I would point to two important solutions that may help: life-cycle analysis and regulatory codes.

Life-Cycle Analysis

To examine the issue of first-cost bias, we need to introduce the crucial topic of *life-cycle analysis*. While not familiar to many homeowners, it is increasingly important in the analysis of sustainability, such as in economics, architecture, engineering, and other fields. Let us see how it works.

Going back to the insulation example, the analysis would start with information on the insulating properties of the two designs. Calculations would include not just the first cost but the future costs.

Table 12-1 shows how to estimate the life-cycle cost of a 20-year investment. Start by listing the outlays in each year. The first year,

TABLE 12-1. Example of life-cycle analysis of insulation investment

Year	Low first cost	Energy efficient	Difference
0	*5000*	*7000*	*2000*
1	900	500	−400
2	900	500	−400
3	. . .	. . .	. . .
4	. . .	. . .	. . .
6	. . .	. . .	. . .
7	. . .	. . .	. . .
8	. . .	. . .	. . .
9	. . .	. . .	. . .
10	. . .	. . .	. . .
11	. . .	. . .	. . .
12	. . .	. . .	. . .
13	. . .	. . .	. . .
14	. . .	. . .	. . .
15	. . .	. . .	. . .
16	. . .	. . .	. . .
17	900	500	−400
18	900	500	−400
19	900	500	−400
20	900	500	−400
Present value	**16,216**	**13,231**	**−2,985**
Payback period (years)			5
Rate of return			19%
Present value at 5%/year			$2,985

in bold italics, is the capital outlay. For people focusing on first cost, the decision ends right there, and the inexpensive version is selected. However, a correct life-cycle analysis would also consider the operating costs, which are in the rows for years 1 through 20.

Finance specialists use three different methods for selecting investments. The simplest approach is the *payback period*, which tells you how long it takes to recoup the additional first cost. In this case you get your investment back in five years, which is reasonably short.

The payback period does not work well if the flows are uneven over time, and most financial specialists prefer the next two. The second approach is the internal rate of return on the investment. This calculates the yield on the investment averaged over the period—roughly the net income divided by the investment. A spreadsheet will show that the return is 19% per year. The return approach is useful because you can compare it with other potential investments. So, for example, if you have $2,000 sitting in a savings account earning 5% per year and you do not need it for a few years, putting that money into insulation would be a wise move.

A final concept is the present value of an investment, mentioned above. This tells you the value in today's dollars of the inflows and outflows of costs and savings. The present value of the expensive insulation is almost $3,000 more than the cheap one if we discount at 5% per year. So making the investment is the equivalent of a $3,000 windfall.

As a final note, return to the bias of overdiscounting. Suppose that people suffer from this bias and discount future benefits not by the correct 5% discount rate but at the hyperbolic 20% rate. In that case, the energy-efficient investment has a lower present value.

Which of these three criteria should you use in making life-cycle investments? The key point is simply that you should definitely use at least one of them because they involve moving beyond first-cost to life-cycle analysis. For most investments, they give similar answers. Either it is a good investment with a short payback period, a high return, and a positive present value—or the opposite. But unless you do the life-cycle analysis, you may find yourself with a lot of inexpensive capital that costs a fortune to operate.

Some Technical Details on Life-Cycle Analysis

For those interested in doing a life-cycle analysis, three further points will be helpful. These relate to inflation, taxes, and risk.

On inflation, the example in table 12-1 assumes zero inflation. A correct analysis would need to incorporate the trend in fuel or other future costs along with the appropriate nominal or money interest rates.

Taxes are more complicated. Energy-saving investments may have further benefits if the government provides subsidies to these investments. For example, in 2016 the following qualified for a 30% federal income tax credit: solar panels, solar-powered water heaters, geothermal heat pumps, small wind energy systems, and fuel cells. Additionally, energy use is not a deductible expense for homeowners. So reducing energy use increases posttax income. If investment income is taxed, then there may be a further advantage from energy-efficient investments.

A final complication is risk. If we take the insulation example, there are several risks that need to be considered. The most important is that the savings are likely to differ from the engineering estimates. Actually, it has been a common observation that energy savings from engineering studies of conservation are overestimated. There are also weather risks—perhaps winters will be warmer than average. A further risk is from fire or other damage since the insurance company is unlikely to reimburse for expensive insulation. Or there might be a major shock to housing markets, such as that seen after 2006, and the value of the investment declines. Finally, perhaps you will sell the house before 20 years pass, and the next owner will suffer from first-cost bias and not want to pay for the additional insulation.

However, let us not lose sight of the key point. In making an investment, make sure to use life-cycle analysis. Include all costs, not just the first costs. Such calculations are particularly important for long-lived investments (like structures) and those that have high future operating costs (such as energy-using equipment).

Regulatory Approaches to Behavioral Problems

If you find yourself a bit fuzzy on the details of life-cycle analysis, you are not alone. It is not part of the curriculum of most schools or of daily life. Whenever I do a real-world life-cycle calculation—say, for insulating my house—it is a nightmare because the data are not available, and I get different recommendations from different specialists. This is not a recommendation to ignore life-cycle analysis, particularly for big-dollar items, but rather to recognize their difficulties.

Once we recognize the difficulties of life-cycle analysis, they suggest that the government has an important role here in regulatory approaches. Suppose that people do indeed systematically undervalue future energy costs. Or perhaps people have too-high discount rates. Or buyers have poor information. Or builders can hide their shoddy work in the walls of the houses they build. To combat these failures, governments can require energy-saving designs for housing and appliances.

Over the last half century, governments have increasingly regulated the efficiency of energy-using capital. Regulations cover automobiles, structures (such as single-family homes), and appliances (such as refrigerators and air conditioners). Efficiently designed capital reduces the need for complicated life-cycle analyses. The idea is that the government can do a life-cycle analysis and pick a lower-bound standard that cuts off the inefficient designs.

Building codes are the major way that governments can affect housing designs. Regulating buildings is important because they are extremely long-lived (my house was built in 1905). Moreover, few structures are designed by architects. However, all buildings are governed by state or local building codes. So the most effective way to introduce Green architecture is through improving building codes.

Here is a way to think about efficiency standards and building codes. Suppose that people (including manufacturers and builders) have first-cost bias and tend to underinvest in various aspects of sound design, Green or otherwise. The codes are ways of preventing the least efficient designs. They are like speed limits in that they keep the most dangerous and inefficient cars, appliances, and houses off the road.

13

Green Political Theory

We now turn to the questions of Green politics—or how our political systems deal with pollution and other harmful externalities. This chapter focuses on political theory, while the next two chapters will look at examples.

Our analysis has emphasized the core concepts of Green phenomena that involve externalities or spillovers. These are activities by firms or individuals that have direct nonmarket effects on others. Externalities operate at different levels and must therefore be managed by different mechanisms.

Mechanisms for Personal Spillovers

We begin the discussion of managing spillovers with the example of *personal spillovers*. The most pervasive spillovers occur in families. This point may be surprising, but most people learn about dealing with conflicting objectives at home.

Suppose you are a nonsmoker and marry a smoker. Both of you know that secondhand smoke is dangerous. What do you do? You will need to negotiate an arrangement with your spouse. You might talk calmly, yell and scream, or take your leave. Whatever your approach, these negotiations are personal and do not involve government coercion. For most cases of conflicting personal objectives, negotiating is the standard solution, whether among families, neighbors, or coworkers. The state becomes involved only in extreme cases, such as physical violence and child neglect.

Societal Spillovers and the Theory of Efficient Pollution

Other spillovers, particularly the ones emphasized in this book, involve *impersonal spillovers*. Examples range from the local and minor, such as street litter or traffic noise, to the global and major, such as climate change and virulent pandemics.

Before turning to Green politics, it will be helpful to address two questions: the theory of efficient internalization of pollution and other externalities: and the tools that can be used to pursue efficient Green policies. The idea was introduced in the earlier discussion of Green efficiency but bears further elaboration in the context of politics.

An amusing and instructive way to think about efficient pollution is to recall the story of Goldilocks. Goldilocks tastes different bowls of porridge in the house of the bears. One is too hot, the second is too cold, but the third is just right, and she eats it all up.

Environmental policy also follows the Goldilocks principle. Optimal regulation comes when the costs and benefits are neither too strong, nor too weak, but just right and appropriately balanced.

> An unregulated market economy will produce too much pollution. In an unregulated state, the social benefit of additional abatement (marginal benefit) exceeds the social cost of additional abatement (marginal cost). Efficiency requires that the marginal social benefit equals marginal social abatement cost.

What is the logic? Firms will generally spend little on abatement in an unregulated market. At zero abatement, reducing pollution has a large benefit and costs very little, so the net benefits of reduction would be high. At the other extreme, it is inefficient to reduce every last speck of pollution, because the costs of going so far would outweigh the benefits. Rather, efficiency comes when an extra unit of cost of pollution reduction is balanced by the total social benefit.

Tools for Efficient Policy toward Externalities

In light of the condition for efficient management of externalities, what tools can governments use to combat impersonal externalities? The most visible activities are government antipollution programs, which induce firms to correct externalities using either direct controls or financial incentives. More subtle approaches use enhanced property rights, which give the private sector instruments for negotiating efficient solutions.[1]

Government Programs

For almost all serious externalities such as pollution, health, and safety, governments rely on *direct regulatory controls*, called social regulations. An important case was the 1970 Clean Air Act, which mandated that automobiles reduce allowable emissions of three major pollutants. For example, standards in cars required a reduction in emissions of at least 90% for carbon monoxide.

These are sometimes called *command-and-control* approaches because they parallel the structure of military decisions. Under military command-and-control, a general recognizes what needs to be done and sees to it that appropriate actions are taken. The general says, "Do this and don't do that." In environmental regulation, similarly, the government commands firms to take certain steps, such as put a catalytic converter in a car to reduce tailpipe emissions. In such direct regulations, it is presumed that the government knows the best technology, and the firms need only to follow the orders, like good soldiers.

Standards are widely used and have a large and sophisticated body of regulations to guide enforcement. In the face of shifting political tides, these standards have proved durable. However, they suffer from the disadvantage of pervasive economic inefficiency. Recall the fundamental requirement of efficient environmental regulation, that "marginal social benefit equals marginal social abatement cost."

In reality, most regulations are set without comparisons of marginal costs and marginal benefits. Moreover, the government seldom knows the best technology to meet emissions goals, and firms can attain the required reductions at lower cost, sometimes far lower costs. Additionally, different firms may have different production structures, which means that some can economically reduce emissions while others may economically do very little.

One example of inefficiency concerns location. Most regulations in the United States apply equally in all areas. Yet the damages from pollution are much higher in densely populated cities than in sparsely populated rural areas. Moreover, regulations often apply different standards to different sources. SUVs have less stringent fuel-efficiency standards than sedans, for example. Studies have found that command-and-control regulations cost significantly more than what is necessary to reach their environmental goals.

To avoid the pitfalls of direct control, many economists advocate *market-type regulations*. In essence, market-type regulations harness the market to fix its own failures.

One approach is the use of emissions fees, which would require that firms pay a tax per unit of pollution. For example, a carbon tax might be $40 per ton of carbon dioxide (CO_2) emissions as an estimate of the marginal damages of emissions. An appropriate emissions fee internalizes the externality by making the firm pay the social costs of its activities. If the emissions fee is set at the marginal social damage, profit-minded firms would be led, as if by a mended invisible hand, to the efficient point where marginal social costs and marginal social benefits of pollution are equal.

Private Approaches

Most people naturally think that market failures associated with pollution and other externalities require government intervention. Legal scholars have shown that strong property rights can sometimes substitute for government actions. This point related to the first pillar of the well-managed society, which is the development of a body of law that defines property rights and contracts so that people can interact in ways that ensure reliable transactions and the fair and efficient adjudication of disputes.

One private-sector approach relies upon *liability laws* rather than direct government regulations. Liability laws internalize externalities by making the injurer strictly liable for damages caused.[2] While liability rules are, in principle, an attractive means of internalizing the nonmarket costs of production, they are limited in practice. They usually involve high litigation costs, which add an additional cost to the original externality. In addition, many damages cannot be litigated because of incomplete property rights (such as those involving clean air), or because of the large number of companies that contribute to the externality (as in the case of chemicals flowing into a stream), or because of legal restrictions in certain areas (such as limitations on class-action suits).

A second private approach relies upon *strong property rights and negotiations* among parties. This approach was developed by the University of Chicago's Ronald Coase, who showed that voluntary negotiations among the affected parties can sometimes lead to an efficient outcome. It is often called the *Coase theorem.*

For example, suppose that I am a farmer using fertilizers that flow downstream and kill the fish in your pond. If your fish business is sufficiently profitable, you may try to induce me to reduce my fertilizer use. In other words, if there is a net profit to be made from reorganizing our joint operations, we have a powerful incentive to work together and agree on the efficient level of fertilizer runoff. You can pay me to stop polluting and still come out ahead. Moreover, this incentive would exist without any government antipollution program.

The Coase theorem is a useful reminder of the power of private bargaining even for impersonal externalities. However, there are many instances where it would not apply. For example, when property rights are poorly defined (as with clean air or climate damages) or when transaction costs are high (in cases with many parties or large uncertainties), it may not be possible to reach a bargain quickly and efficiently.

We see then that there are many approaches to addressing externalities, and care needs to be taken to fit the remedy to the problem at hand. However, at a deeper level, all involve governmental action. The action may be regulation or taxation, both of which are collective acts. However, even defining liability rules and the laws of property are political acts. For example, a country might define a "right to pollute" and then allow markets to arise to buy and sell those rights.

The key point here is that for impersonal spillovers—those that pervade the economy and society—governments are in charge. People may cough and die, firms may prosper or fail, species may disappear, and lakes may catch on fire. But until governments, through the appropriate mechanisms, take steps to control the polluting causes, the dangerous conditions will continue.

Politics as Aggregation of Individual Preferences

We have emphasized the importance of Green federalism, which recognizes that the solution to conflicts over Green policies will necessarily involve different institutions and decision processes. Some are at the individual level, some involve businesses, but the most significant externalities require actions by governments. So here is where *Green politics* enters.

Politics refers to decisions taken by the polity, or people acting collectively, usually by governments. Collective action is required to deal with important externalities such as air pollution, climate change, generation of fundamental knowledge, and the provision of much physical and intellectual infrastructure. These actions are ones in which governments decide for the population by raising revenues, issuing regulations, taxing undesirable activities and subsidizing desirable ones, determining property rights, and setting liability laws.

We can frame politics as a way to aggregate individual preferences (see figure 13-1). Suppose there is an issue, such as dealing with sulfur pollution or protecting Yellowstone National Park. People have

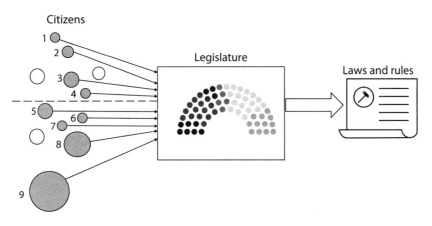

FIGURE 13-1. Up or down? The process by which political systems aggregate preferences of individual citizens

views on the question—some informed, some uninformed. Then, they express their views and influence the outcome. Some citizens do not care and do not vote (these being the hollow circles in figure 13-1). Others have disproportionately large influence through campaign contributions or large entertainment audiences.

Political theory often holds that decisions are made by "the median voter." So the person whose view is in the middle, between both sides, often becomes the decider in elections. Take figure 13-1 as an example, where the issue is UP versus DOWN. This chart shows Voter 5 as the median voter because she is in the middle of the spectrum. The median voter could be crucial in direct voting, as when California had a referendum on hazardous substances (Proposition 65) or when Britain voted on Brexit.

In most circumstances, political decisions take place at a distance from citizens, which means citizens choose legislators who vote on their behalf. Suppose voters can vote for either an UP Party or a DOWN Party. The people above the dashed line in figure 13-1 support the UP Party, while those below are DOWN Party advocates. The DOWN Party gets more votes and controls the legislature. Because the DOWN Party controls the outcome, the median voter moves from Voter 5 to Voter 7, who is the median voter in the DOWN party.

Alternatively, perhaps money influences legislators, which in turn influences the election. Suppose the size of the circle represents dollar influence. If we take the median dollar, that now moves to Voter 8, who is not representative of the party or of the larger population.

Figure 13-1 also shows the effect of "polarized" views and parties, leading to an interesting twist in Green politics. If there were a swing in voter sentiments, the UP Party might gain support because Voter 5 would swing from the DOWN Party to the UP Party. With such a minor swing in sentiment, there is a sharp swing in legislative politics because the median voter of the party in power shifts from Voter 7 to Voter 3.

Additionally, polarization theory suggests that unless the system is stabilized by some institutional mechanism, decisions might swing wildly from one wing to the other as elections sweep one party into power, replacing the incumbent party.

In the United States, many features of institutional inertia prevent wild swings in actual policies. For example, people who serve in the federal judiciary have a lifetime tenure, while members of the Senate serve six-year terms. The high rate of incumbency means that many members of Congress stay for decades once they are established.

Furthermore, the U.S. legal structure produces much inertia because overturning a law requires an act of Congress, which means one party must have control of both houses of Congress and the White House to enact major changes. Indeed, even majorities everywhere are not enough because the Senate requires a supermajority of 60% for most laws. Many other countries also have mechanisms that impose stability, preventing drastic changes based on small changes in public opinion.

Cycles in Environmental Policy?

The United States has evolved a regulatory structure that moves very slowly but is highly stable. Though recognition of the lethal aspects of air pollution can be dated back to the Great London Smog of 1952, it took almost two decades for the recognition of

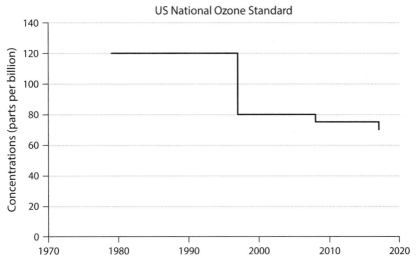

FIGURE 13-2. National ambient air standard, ozone

the health impacts of smog to lead to federal laws and regulations in the United States.

However, once in place, the laws and regulations remained in place through many administrations. Even when new administrations entered who wanted to turn back the clock (as happened during the Reagan period from 1981 to 1988) or tighten rules quickly (as during the Obama period from 2009 to 2016), the inertia of the legislative and regulatory systems meant that both backward and forward changes were slow.

One way to see the stability of the regulatory system is to examine the air quality standards of major air pollutants. Figure 13-2 shows the ambient air standard for ozone, one of the most important and expensive pollutants to control. This standard has never been loosened since it was first promulgated in the 1970s, and the standards were actually tightened by both Democratic and Republican administrations.

How can we understand the stability of environmental rules in the United States? In part, it comes from strong public support. Equally important, though, is how rules are determined through a lengthy set of procedures, informal rulemaking, which requires both detailed proposals and the justification of new rules. When an administration attempts to roll back rules without sufficient justification and legal care, the government's case is susceptible to being overturned by the U.S. courts as "arbitrary and capricious."

Environmental specialists point to the stability of rules compared to taxes as an important advantage. Tax laws can be changed quickly when the political winds shift. The most recent example is the Trump administration tax cuts of 2017, when a massive change in the tax law was written and passed in a few weeks. Environmentalists worry that pollution taxes might be subject to the political guillotine if critics assume power and gain majorities in the major legislative and executive branches.

Is There a Bias in Environmental Laws and Rules?

Notwithstanding the stability of U.S. environmental laws, many people are suspicious that the laws and rules are biased in favor of the "moneyed interests"—the large dollar votes in figure 13-1. Why might this be the case, and what is the evidence?

The simplest approach is what we used above, where politics reflect the central tendency of public views. Suppose that environmental policy, such as limiting pollutants that lead to deadly smog, has costs and benefits distributed among companies and the public. Figure 13-3 uses the same apparatus as figure 13-1. But here we have concentrated and large polluting interests at the bottom, these perhaps being chemical companies owned by rich oligarchs. Producers are few, but their costs are high, concentrated in one or two industries, and billions of dollars of profits are at stake.

However, there are also many people who are suffering ill health from pollution of airborne chemicals. They are represented by the dispersed and uncoordinated tiny circles at the top. Although there are many ill people, they have little political power because they have limited knowledge, are uncoordinated and widely dispersed, and each have only a little to gain from their individual actions.

If we add up the costs and benefits, the net benefits of reducing pollution (benefits minus costs) are positive, indicating that

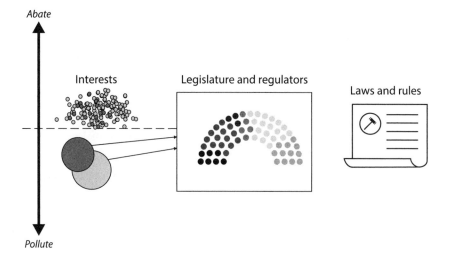

FIGURE 13-3. Concentrated interests (*two large brown circles*) outweigh dispersed and uncoordinated interests (*army of small green circles*)

an environmental policy is in the average interest. However, the average interest often does not prevail. The concentrated brown circles in figure 13-3 organize, hire lobbies, talk to federal and state legislatures, give large campaign contributions, and are critical in close elections.

Just this explanation was provided in a theory of collective action developed by Mancur Olson. It holds that the few can effectively organize to obtain their interests, while the many have insufficient incentives to get organized and represented effectively.[3]

We can use the example of steel tariffs to illustrate Olson's theory. Steel tariffs have been imposed and removed for decades, most recently in 2018. Steel tariffs benefit U.S. steel producers but impose large costs on steel consumers, such as manufacturers and purchasers of automobiles, appliances, and pipelines.

Here is where Olson comes on the stage. The steel industry has its own special lobby and lawyers to argue for tariffs. By contrast, the millions of consumers who use steel products in America have no serious counterweight to the steel industry. Given the imbalance of persuasive powers, it is unsurprising that the steel industry has often prevailed so in its lobbying for tariffs.

Political theory is instructive but also leaves too many possibilities. Should we believe in the median voter theory? If we think the median voter is pivotal, should that be the median voter of all voters, or of the party in power, or of the median dollar? Perhaps the groups with the most cohesive and powerful lobbying arm will determine the outcome. Alternatively, will the persuasive power of public health experts hold sway when they present persuasive evidence on the dangers of smoking or ozone or sulfur? The next chapter examines empirical evidence on Green politics to determine whether there are general patterns to the answers.

Green Politics in Practice

The last chapter described standard approaches to understanding the politics of the environment. This chapter looks at empirical analyses for lessons. It begins with a broad view of major factors such as economic growth and democracy and then examines some areas of particular importance.

Democracy and the Environment

Begin with the impacts of broad political forces on environmental quality. Perhaps the most important question is the role of "democracy" on the environment.

Democracy seems a nebulous term, but political scientists have developed quantitative measures of the state of democracy or autocracy in countries. For example, the Polity project measures democracy using three major factors: One is the presence of institutions and procedures through which citizens can choose their policies and leaders. A second measure is institutional constraints on the exercise of power by the government. The third is the guarantee of civil liberties to all citizens. At one extreme are the full democracies (score = 10) like the United States, Canada, and Germany. At

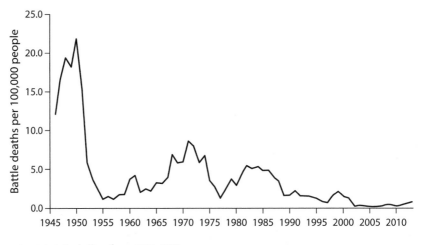

FIGURE 14-1. Lethality of war, 1946–2013

the opposite extreme (score = minus 10) are repressive regimes like North Korea and Saudi Arabia.

What are the findings about the effects of democracy on the environment? Actually, there is virtually no empirical literature on the subject. We can find some isolated studies here and there, but it is largely uncharted territory. The present chapter pulls some of the strands together, but much more remains to be done.

Political scientists have found several features of democratic societies. One of the most important is that democracies are less prone to fight wars against other democracies than are other forms of government. This tendency, sometimes called the Kantian peace after the philosopher who first proposed it, has been a robust finding for many years. You might think that the Kantian peace is far removed from environmental concerns, but in fact war is the sworn enemy of the environment. Indeed, the likely worst outcome for both humans and the globe would be a major nuclear war and nuclear winter.

Figure 14-1 shows the decline in the lethality of war since 1945.[1] The declining prevalence of battle deaths and major wars over the last seven decades ranks as one of the major contributors of democracy to the environment.

The Kuznets Environmental Curve

Another key factor in the development of environmental improvements is economic development. Rich countries tend to be more democratic, and they can also afford strong environmental policies. There is much interest, therefore, in the interaction of wealth and the environment.

One common approach is a theory known as the Kuznets environmental curve, or KEC. This is, roughly, that "the environment gets worse before it gets better." More precisely, the hypothesis holds that the amount of pollution increases in the early stages of economic development with the rise of industry and then declines with higher incomes as services become increasingly important.

The evidence on the KEC is mixed, as we will see. One interesting example is carbon dioxide, CO_2, which is a useful indicator because it is well measured in virtually all countries and is an important contributor to climate change. Figure 14-2 shows the relationship between per capita gross domestic product (GDP) and carbon intensity (CO_2 emissions per unit of output). It is clear that carbon intensity increases for countries up to an income of about $15,000 and then declines. The decline occurs primarily because of the composition of output (moving from agriculture to industry to services). Figure 14-2 shows a hopeful curve for the future, although it will do little to slow CO_2 emissions in the coming years.

While the KEC in figure 14-2 looks convincingly dome shaped, other indicators give different answers. For example, if we were to plot total CO_2 emissions on the vertical axis of figure 14-2, we would find that the curve rises at all levels.

Yet another trend comes for local pollutants such as fine particulate matter (PM2.5, particles with a diameter of less than 2.5 micrometers). PM comes from many sources, but the most important from the viewpoint of pollution control are the emissions from the combustion of coal. The present analysis looks at concentrations rather than emissions because concentrations are the important ingredient in health damages.

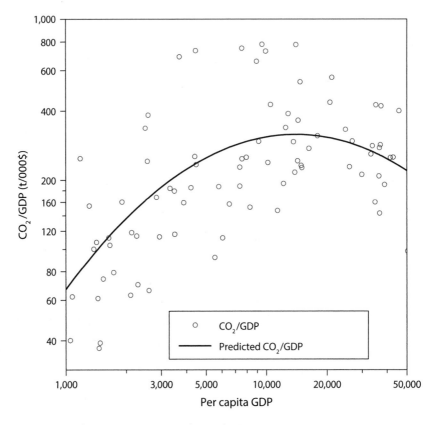

FIGURE 14-2. The Kuznets environmental curve for CO_2 in 2000

I discuss the U.S. experience with PM regulations later in the chapter, but we can examine the experience around the world in the present context. Recent data suggest that PM concentrations do not follow the Kuznets curve. Rather, concentrations decline at every level of output for major countries from 1990 to the present. For the year 2010, every doubling of per capita output was associated with a 25% decline in PM2.5. The income-pollution relationship can be seen by grouping countries at the top and bottom. The twenty poorest countries in 2010 had average PM2.5 concentrations of 41 μg/m³ (micrograms per cubic meter), while the richest twenty had average concentrations of 14 μg/m³.

A second relationship is between emissions and democracy. We might think that democracies should be cleaner because they are more responsive to popular will, people are more informed and can express

their views, and democratic governments are more likely to cooper-
ate with other countries on international environmental problems.

On the whole, the evidence supports the pro-environment
nature of democratic institutions. It is generally found that well-
established democracies have lower pollution. As one study concluded, "Democracy can help improve the environment, but only
if given time to promote mechanisms of accountability, facilitate
information, foster associational life, spur international cooperation,
and promote institutional development."[2]

Returning to the example of PM, figure 14-3 shows the impact of
democracy scores on PM concentrations for the 80 largest countries
with data. It is clear that democracies have cleaner air. For example,
a complete democracy is estimated to have about 45% less pollution
than a complete autocracy (correcting for other factors). Countries
with higher democracy scores also showed larger declines in PM
concentrations over the 1990–2010 period than autocracies. While
the size of the impact is impressive, the statistical association is sensitive to the sample and specification.

The conclusion about the importance of democratic institutions
reinforces the observations from the last chapter. The earlier discussion suggested that the institutional structure in the United States
is slow to react to environmental challenges but is also durable and
able to resist abrupt and temporary shifts in the political winds. The
experience of other countries reinforces this observation, indicating
the power of democratic forces on the establishment and durability
of strong environmental policies.

Politics, Distributional Politics, and Environmental Policies

Empirical studies of the politics of environmental regulation demonstrate the importance of the two factors emphasized in the last
chapter. We saw the tension between social welfare and the power
of interest groups. Sometimes, concentrated interests (such as the
steel or oil industries with a small number of large and wealthy firms)
outweigh the dispersed interests (of hundreds of millions of oil consumers with little knowledge or resources). Still, we might want to

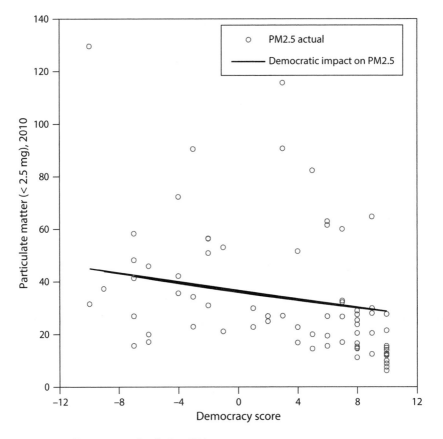

FIGURE 14-3. Democracy and pollution, 2010
This graph shows the association of countries' concentrations of particulate matter (PM2.5) with their democracy scores. The line is the prediction from the statistical regression, while the dots are for individual countries.
Source: data on global mortality from pollution are from Michael Brauer, Greg Freedman, Joseph Frostad, Aaron Van Donkelaar, Randall V. Martin, Frank Dentener, Rita van Dingenen et al., "Ambient Air Pollution Exposure Estimation for the Global Burden of Disease 2013," *Environmental Science and Technology* 50, no. 1 (2016): 79–88, doi:10.1021/acs.est.5b03709; the World Bank for output and population; and the Polity website for democracy scores, https://www.systemicpeace.org/polityproject.html.

know just how much the interests of the concentrated and moneyed dominate those of the public, who tend to be more concerned with their daily lives. In short, does environmental policy approximate social welfare or fall far short of it?

There is no overall measure of the efficiency of environmental policy. The balance of this chapter takes three examples to illustrate how Olson's theory of collective choice has played out historically on distributional politics as well as the environment: the tariff, sulfur politics, and climate change.

Tariffs

Begin with the tariff. This may seem far removed from environmental policies. However, it is a classic area for the study of the conflict between narrow and broader interests. Tariffs are taxes levied on narrow groups of imported products (such as steel or lumber). They tend to favor a narrow group (domestic firms and workers in the industry) and harm the broad group of consumers.

An example will illustrate the point. Suppose the United States puts a 10% tariff on imported steel. Here are the current numbers, simplified but accurate enough for the point. The United States consumes about 100 million tons of steel costing about $100 billion, of which 30% is imported. If tariffs raise steel prices by 10%, consumers will pay an extra $10 billion. Of this total, about $7 billion would go to the steel industry, perhaps divided among 70 firms, each having an additional $100 million in profits. Therefore, the higher profits among steel firms would be the concentrated gains. By contrast, a $10 billion loss would be divided up among 330 million consumers, with the average loss being $33 per person. While simplified, steel tariffs are a fine example of concentrated gains and dispersed losses.

James Madison clearly described the nature of tariff politics in *Federalist Paper 10*:[3]

> A landed interest, a manufacturing interest, a mercantile interest, a moneyed interest, with many lesser interests, grow up of necessity in civilized nations, and divide them into different classes, actuated by different sentiments and views. The regulation of these various and interfering interests forms the principal task of modern legislation. . . . "Shall domestic manufactures be encouraged, and in what degree, by restrictions on foreign manufactures?" are questions which would be differently decided by the landed and the manufacturing classes, and probably by neither with a sole regard to justice and the public good.

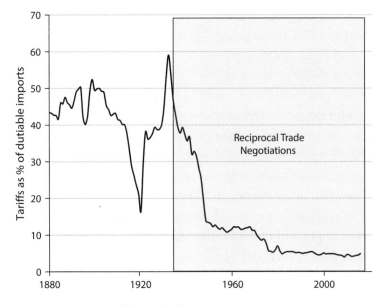

FIGURE 14-4. Average tariff rate, United States, 1880–2015

The history of the tariff in the United States showed the wisdom of Madison's observations. Because of the play of sectional and industrial interests, the United States has been a high-tariff country for most of its history. For the period from 1880 to 1930, the average U.S. tariff rate was almost 40%, as shown in figure 14-4.

Tariffs, sometimes called economic protection, have largely focused on manufacturing, with high tariffs shielding domestic workers and firms from imports. Manufacturing was located in the northeast and north-central United States, roughly from Maine through Illinois. This region has historically been the center of political support for high manufacturing tariffs (for example, the tariffs of 1828 and 1929). It is striking that the protectionist rhetoric of the 2016 Trump campaign was most successful in the same north-central states that had been the source of support of tariffs since the early years of the republic.

The seesaw patterns of the tariff rate up to 1929, shown in figure 14-4, largely reflected the political fortunes of the two major parties in Congress.[4] The implication for understanding Green politics is that it was, as Madison foresaw, primarily the result of the battle of concentrated regional interests that determined the outcome, not decisions on the efficient design of a fiscal structure.

The tariff is a fine example of the imbalance of consumer and producer interests. For decades, concentrated producer interests won out, and tariffs were high, as figure 14-4 shows. Why and how did this end? Two fundamental changes occurred in the early twentieth century. The first was the introduction of the income tax in 1914. This reduced the fiscal necessity of the tariff and was critical for financing wars and the rise of the welfare state after World War I.

The second important step was the emergence of *reciprocity*—the recognition that countries would need to reduce their tariffs if they wanted tariff reductions abroad to promote exports. Reciprocity in the United States became particularly important as the country moved from a net importer of manufacturing goods (up to 1910) to a major exporter after World War I.

The change from noncooperative high tariffs (through 1929) to a cooperative, negotiation-based, reciprocal approach took place almost by accident during the early years of the Franklin D. Roosevelt administration (1933–1945). Roosevelt believed (incorrectly) that high tariffs were a major determinant of the length and depth of the Great Depression, but he had no clear view on tariff policy. However, Roosevelt's secretary of state, Cordell Hull, was a firm advocate of reducing trade barriers through cooperation and negotiations. In 1933 he wrote in words that should be read carefully today:[5]

> Many years of disastrous experience, resulting in colossal and incalculable losses and injuries, utterly discredit the narrow and blind policy of extreme economic isolation. . . . After long and careful deliberation, I decided to announce and work for a broad policy of removing or lowering all excessive barriers to international trade, exchange and finance of whatsoever kind, and to adopt commercial policies that would make possible the development of vastly increased trade among nations. This part of my proposal was based on a conviction that such liberal commercial policies and the development of the volume

of commerce would constitute an essential foundation of any peace structure that civilized nations might erect following the [first world] war.

Starting with the Reciprocal Trade Agreements Act of 1934 and going through several rounds of negotiations, the United States and other nations progressively dismantled their tariff and nontariff protectionist structures. The rise of globalization came in large part because of the spirit and substance of international cooperation to reduce barriers to trade and finance.

Recent history is a reminder that nations take one step backward for every two steps forward. The administration of Donald Trump has used tariffs and protectionist bluster for political purposes, alienating allies and undermining the international trading regime. Small trade wars have resulted, but as in military wars, they have produced few true victors.

As of 2020 the future of the international trading system is cloudy. But the lesson for Green politics is instructive. In a historical period when regional and industrial interests dominated, the few won out over the interests of the many. The iron grip of narrow interests on the tariff gradually loosened because of a view in which *mutual cooperation and removal of the powers of special interests* came to dominate trade policy. Special interests were vanquished by cooperative policies that took the distributional politics of the tariff out of the hands of Congress.

Sulfur Politics

The politics of air pollution, especially the oxides of sulfur (SO_x), is perhaps the most consequential of all major Green domestic concerns (climate change being the major long-run global issue). Sulfur dioxide (SO_2) is an emission that arises primarily from the burning of coal and other fossil fuels but has other sources such as mining and wood-burning stoves. SO_2 is harmful, but it becomes particularly dangerous when combined with other compounds to form small particles or particulate matter (PM). The sulfur oxides and PM lead to serious health effects in humans, as well as damage to ecosystems.

An interesting feature of sulfur politics was the focus of early environmental activism on "acid rain." Acid rain occurs when sulfur oxides (which are acidic) raise the acidity of lakes and forests and injure these ecosystems. Among the adverse impacts are damages to fish and other life in freshwater, harm to soils through acidification, and indirect effects on forests. The first major government policies to control sulfur emissions were motivated by the dangers of the ominous-sounding acid rain in the late 1970s and beyond.

The politics of sulfur (and more generally of environmental policies) are well illustrated by the evolution of the debate about acid rain. When the conservative Reagan administration took office in 1981, one of their targets was reducing environmental regulation, and they quickly focused on the proposed regulations on acid rain. Buttressed by external, probusiness "scientists," the administration argued that the science behind acid rain was too unsettled to form the basis of costly regulations. There were, according to the Reagan administration, uncertainties about cause and effect as well as the mechanisms driving acid rain. Conflicting testimony and experts tended to reinforce the uncertainties.

On a parallel track, public health scientists were studying the impact of air pollution on human health. The association between air pollution (smoke and SO_2 emissions) and health gradually emerged, and by the 1970s public health scientists had concrete evidence of the linkage. The quantitative exposure-response relationship was firmly documented during the 1980s and 1990s.

The work of public-health specialists finds that the global health effects of air pollution are extremely large. It is estimated that more than 4 million premature deaths globally in 2015 were linked to air pollution, of which half were in China and India. The pollutants most strongly linked are PM, ozone, nitrogen dioxide, and SO_2 (sulfur dioxide). If we examine the mortality from particulates alone, deaths in the United States were estimated to peak at 90,000 in 1980, declining to 68,000 by 2006 and to 36,000 by 2016.[6]

Regulation of SO_2 and other air pollution took place at the local, state, and federal levels, but this discussion will focus on federal policies. The major statutes were the Clean Air Act of 1970 and amendments in 1977 and 1990. Each of these allowed new approaches and a tightening of standards.

The 1990 amendments were one of the landmarks of environmental law in introducing an approach known as *tradeable permits*, or *cap-and-trade*. Under this radically new approach, the nation would cap the amount of annual *national* sulfur pollution. The government would allocate emissions permits to firms, who would own them. Additionally, firms could buy and sell allowances as needed for their operations. This system ensures that the overall level of pollution is attained at minimum cost.

The trends in major air pollutants have been impressive in the United States. Take as an example the most damaging source: industrial SO_2 emissions declined from a peak of 31 million tons per year in 1970 (when the Clean Air Act was enacted) to around 2.7 million tons in 2016.

The important question for Green politics is how well political systems have responded to the risks of sulfur pollution. For this discussion, I will focus on the United States, looking at the period from 1990 to 2016. The year 1990 was a turning point with the United States' passage of the law that enabled the trading of sulfur-emissions permits. Table 14-1 breaks the last quarter century into four periods (1990, 2000, 2006, and 2016) and shows the growth rates over this period.

Focus on the last column, which shows concentrations of the most damaging pollution (small particles or PM2.5) relative to electricity generation. This shows a sharp downward trend, averaging more than 4% per year in the concentrations per unit of electricity. The most impressive change was the decline in SO_2 emissions, with total emissions falling from 31.1 million tons in 1970 to 13.1 million in 2006 to 2.3 million tons in 2016. The decline in sulfur emissions had many sources but primarily resulted from increasingly tight federal regulations. The system of tradable allowances was particularly effective in the period after 1990 while strong regulations on toxic

TABLE 14-1. Major factors determining improving air quality

Period	Electricity Generation	Coal Consumption	SO$_2$ Emissions	PM10	PM2.5	PM2.5/ Electricity Generation
	Rate of change in measure (average percent per year)					
1990–2000	2.3%	1.8%	–3.4%	–2.5%	–2.8%	–5.1%
2000–06	1.1%	0.4%	–3.7%	–1.4%	–2.5%	–3.6%
2006–16	0.0%	–4.2%	–15.8%	–1.6%	–4.0%	–4.0%

air emissions in the 2011–2015 period were an additional factor. Yet other forces, shown in table 14-1, included the leveling off of electricity production and the declining use of coal (the major source of sulfur), along with the increased competitiveness of natural gas.

The next question for Green politics, one that is extremely difficult to answer, is whether the overall regulatory stringency in air pollution meets the Goldilocks criterion of not too strong and not too weak—that is, where the costs and benefits are appropriately balanced. How do the marginal social benefits of pollution reduction compare with marginal costs of reduction?

This question of overall stringency was addressed in a study by Nick Muller and colleagues.[7] They estimated that the incremental damage caused by a ton of SO$_2$ emissions in 2010 was about $2,000 per ton of sulfur emitted. The average price of emissions permits in that year was about $40 per ton. This difference between the price of sulfur emissions and the marginal damage of emissions indicates that sulfur regulations were much too lax.

Moving forward to 2017, the price of emissions had fallen to 6 cents per ton, so emissions were essentially free. The emissions price fell so sharply because actual emissions were low relative to the quantitative regulatory standard. There were no calculations of the marginal benefits of emissions reductions for that year, but recent estimates of marginal damages of sulfur emissions are above $6,000 per ton.[8]

Regulations have succeeded in reducing emissions dramatically— by a factor of ten over the last half century. But the regulatory regime is currently too lax in the sense that the marginal cost of reducing sulfur (its trading price) is far below its marginal benefit in improving public health. As noted above, about 36,000 premature deaths occur annually in the United States from particulates, and further tightening of standards could substantially reduce that number.

Additionally, sulfur regulation illustrates an important problem with the original cap-and-trade regulations. When emissions decline below the quantitative limit (either for reasons related to regulation or for market reasons), then the market price of that pollutant can fall very sharply, even to zero. This was seen as well with the European Trading Scheme (ETS) for CO_2 in its early years. The sharp decline in prices can be prevented by the use of price floors, or even better, as we will see later, by the use of emissions taxes (carbon taxes for CO_2, or sulfur taxes in the present case). The point is that the price of emissions should reflect the marginal damages from an additional unit of emissions, and that will not fall to zero even when emissions decline below the quantity target.

Climate Change Policies

Climate change is a final example we can use to examine how well the political system is responding to a Green challenge. We will discuss the issues involved in climate change in our chapters on global Green. Those chapters will describe how an efficient policy would involve assigning a price to CO_2 and other greenhouse gases that reflects the marginal damages from those emissions. The marginal damage of CO_2 emissions has a special name, the *social cost of carbon*, or SCC. Most calculations are for a *global* SCC, which reflects the damage to all countries. An efficient policy would involve each sector in each country having a *harmonized carbon price* equal to the SCC. Setting the universal global harmonized price at the right level is "all" that is required for an efficient policy for climate change.

At first blush, this requirement seems unbelievably simple. Is it really true that the only requirement for efficient regulation is a harmonized carbon price at the SCC? Yes, and the reason is straightforward. A molecule of CO_2 emissions mixes into the atmosphere with all the other CO_2 molecules. Each molecule is anonymous, so to speak, and has the same impact over the coming years. Therefore, the regulatory price attached to each molecule emitted from every source around the world should be the same.

What is the actual number for the SCC? The calculation is extremely involved and has been made by several modeling efforts. The U.S. government surveyed several studies and estimated that the appropriate global SCC was around $40 per ton of CO_2 in 2020. There are large uncertainties about this estimate, and it almost surely is less than the price that will attain the international objective of limiting temperature increase to 2°C. However, we can focus on this U.S. government number for the present discussion.

The key question for green politics is how the level of emissions reductions, or the effective carbon price, around the world compares with the efficient price. The World Bank estimates that the actual global average price in 2018 was two dollars per ton of CO_2, or about one-twentieth of the SCC.[9] The only region with a carbon price that applies to the entire region is the European Union. Other major countries (China, the United States, and India) have some regional carbon pricing but as of now no national pricing.

Therefore, the bottom line for climate change is that current policies are much weaker than the efficient policy or the policy that would attain current international climate objectives.

Why has climate policy failed so miserably while policies for many other pollutants have succeeded? The reasons for the political failures to meet the needs of climate change policy will be discussed extensively in the analysis of climate change but can be summarized briefly. Some sources are the same as for domestic policies. Strong climate change policies have widely dispersed benefits and concentrated costs. Additionally, many of the benefits come far in the future, and political systems often overdiscount future benefits.

However, the main reason for the policy failure is that climate change is a *global externality*. Countries might be interested in their *national* SCC but have little interest in the *global* cost. As a result, and therefore unlike domestic policies, climate change is subject to international free riding. This is a syndrome in which countries are reluctant to take strong policies because most of the benefits accrue to other countries. We will see in the chapters on global Green that free-riding is a central defect in current global climate-change policies.

Putting all these reasons together, we see that climate policy faces ferocious headwinds. As a result these policies are much too modest to make a serious dent on CO_2 emissions, rising temperatures, and the encroaching oceans.

Conclusions on Green Politics

Among the many findings that emerge from the last two chapters' review of Green politics, three stand out. First, it must be emphasized that many of the challenges of the Green movement can only be met by governmental policies. These range from setting a legal framework in which entities are responsible for their harmful actions to regulatory policies for the most important spillovers like pollution and infectious diseases. This point is a reminder of the need for governments to provide public goods as one of the central pillars of the well-managed society.

A second finding is that environmental policies often lag many years behind scientific findings. Scientists knew of the dangers of tobacco, of sulfur emissions and smog, of climate change, and of pandemics long before effective government actions were taken. This lag occurs partially because governments—even democracies—have great inertia in their actions. Political action requires gathering evidence, weighing interests, overcoming objections, passing laws, devising regulations and enforcement measures, and finally taking actions. The lag also reflects the power of concentrated vested interests in blocking the interests of the uncoordinated and dispersed who are adversely affected. Added to these are hindrances erected when political leaders come under the sway of the moneyed interests and antiscience factions in a society.

A third finding is the need for cooperation and coordination to overcome factional interests on the national stage and free-riding at the international level. In all three case studies discussed here—international trade, sulfur pollution, and climate change—a lack of cooperation impeded progress toward an effective institutional system. For trade and sulfur, cooperation was eventually put in place, and the collective interest was pursued. However, in the case of climate change, free-riding and lack of coordination continue to be the major impediments to effective policies.

15

The Green New Deal

Starting in 2018, the idea of a *Green New Deal* (GND) flashed across the American consciousness. It arose in part because of the determination of the Trump administration to scrap many environmental policies and partly from the recognition of the grave consequences of climate change. It also captured the hearts of the liberal wing of the Democratic Party in the United States. Given its importance— even perhaps for a fleeting time—its ideas are worth a careful review.

The New Deal

The Green New Deal (GND) takes its inspiration from the New Deal of the 1930s. The original New Deal was a set of innovative policies introduced during the Great Depression by Democratic president Franklin D. Roosevelt. By the time Roosevelt took office in March 1933, output had declined 30% from its 1929 peak, and the unemployment rate was 25% of the labor force.[1]

FDR ranks as one of the greatest presidents in America's history. Above all, he is admired by historians for his role as a coalition builder and military leader who led the embattled democracies to victory in World War II. Reviews of the economic policies during the New Deal are less starry-eyed than his political and military record.

What were the central economic components of the New Deal? His efforts to end the Great Depression were largely based on discredited economic reasoning. A first factor was that the New Deal was fundamentally experimental and therefore deeply radical in its willingness to upset societal apple carts. For example, FDR's initial impulses leaned toward fiscal conservatism, reducing fiscal deficits, and raising taxes.

The influence of economists like J. M. Keynes, as well as FDR's own instincts, led to a radical revision in FDR's fiscal policies. Federal nondefense spending rose from under 2% of gross domestic product (GDP) at FDR's inauguration to 5% of GDP on the eve of World War II. Indeed, federal investment as a share of the economy reached its all-time peak during the late 1930s, and the share of federal spending on goods and services has been declining ever since.[2]

As large as it was, the fiscal stimulus from the New Deal was completely inadequate for the times. An additional ingredient in the recovery came when FDR took important steps to demolish the gold standard as a "golden fetters," in Keynes's phrase. None of these were sufficient. Economic historians have concluded that it was only the massive fiscal stimulus of military spending for World War II that pulled the U.S. economy out of the depression.[3]

The second feature of the New Deal was FDR's willingness to use the full power of the federal government to combat economic ailments. Before 1933 there were relatively few government agencies and little federal government spending. The New Deal period led to a proliferation of programs and agencies. Ones that survived include the Civil Aeronautics Board, the Export-Import Bank, the Federal Communications Commission, the Federal Deposit Insurance Corporation, the Federal Housing Administration, the National Labor Relations Board, the Securities and Exchange Commission, and the Social Security Administration.

In retrospect we find that many of the programs were both well conceived and durable. These include social security, unemployment insurance, deposit insurance on bank deposits, and securities law. Others, such as the National Recovery Administration, disappeared with little to remember when the economy returned to full employment during the 1940s. Today's Green New Deal follows its predecessor in proposing a wide array of new or expanded social and economic programs to meet its goals.

A third feature of the New Deal was its emphasis on reducing economic inequality. Since the Great Depression had reduced incomes and destroyed families and communities, the most important way to reduce poverty and inequality was a return to full employment. Other key policies were Social Security (ending poverty for the elderly), welfare (supplementing incomes for low-income groups), and unemployment insurance. Here again, the GND follows up on unmet needs.

Fourth, important elements in the New Deal of the 1930s led the way in introducing environmental policies of the federal government. The Roosevelt administration pioneered several elements in the area of conservation. An act on soil conservation in 1936 responded to the terrible droughts and storms of the "dust bowl," which turned a huge area from Texas to South Dakota into a wasteland. A second program was the Civilian Conservation Corps (CCC) of 1935. A small army of young men lived in CCC camps, restoring historic sites, killing grasshoppers, and building towers and trails. The CCC undertook much of the planting of forests in the United States. The spirit of the conservation projects of the New Deal predated the environmental movement of the modern era and put little emphasis on externalities and pollution. Rather, the original New Deal made the critical breakthrough of providing legitimacy, as well as the instruments, for national policy to intervene actively in economic affairs that affected both humans and nature.

Looking back at the New Deal today shows that the government can be more than the night watchman; it can serve as a "powerful promoter of society's welfare," in the words of jurist Felix Frankfurter. His reminder is particularly important in this era when many governments, including the United States, have sometimes served the welfare of leaders or parties rather than society and have done so in a manner that misleads and upends the valuable traditions of an open and democratic society.

The Origins of the Green New Deal

While the idea of a Green New Deal surfaced occasionally in the last two decades, it was often adopted as a mechanism to adapt current economic and political structures to meet the challenges of ecological and environmental crises.[4] The origins of the idea of the GND apparently originated in two 2007 columns by *New York Times* columnist Tom Friedman.

> If we are to turn the tide on climate change and end our oil addiction, we need more of everything: solar, wind, hydro, ethanol, biodiesel, clean coal and nuclear power—and conservation. It takes a Green New Deal because to nurture all of these technologies to a point that they really scale would be a huge industrial project.
>
> [We] need a Green New Deal—one in which the government's role is not funding projects, as in the original New Deal, but seeding basic research, providing loan guarantees where needed and setting standards, taxes and incentives.[5]

The first systematic writing originated in Britain, where the New Economics Foundation produced a pamphlet in 2008, *A Green New Deal*.[6] In addition to climate policy and infrastructure, the proposal called for "a 'carbon army' of workers to provide the human resources for a vast environmental reconstruction program." An interesting feature of the early proposals was their emergence during the midst of the 2008–2009 financial crisis and economic downturn. It was in this economic downturn that the emphasis on job creation and economic stimulus formed a central core of the early proposals and got swept into the 2019 proposals.

The notion of a Green New Deal has been in the wind in many circles since launched by Tom Friedman—in speeches from diplomats, legal scholars, environmentalists, and, occasionally, economists.

The Green New Deal of 2018-2020

The Green New Deal took a new turn after the U.S. election of 2018 under the inspiration of a cohort of new and more progressive Democratic members of Congress. A major milestone was reached in February 2019 when Representative Alexandria Ocasio-Cortez introduced a House resolution, Recognizing the Duty of the Federal Government to Create a Green New Deal, with Senator Edward Markey introducing a parallel Senate resolution.

While widely acclaimed by many progressives and environmentalists, the approach was soon caught in a partisan cross fire. Republicans in the House introduced an opposing resolution "expressing the sense of the House of Representatives that the Green New Deal is antithetical to the principles of free-market capitalism and private property rights, is simply a thinly veiled attempt to usher in policies that create a socialist society in America, and is impossible to fully implement."

What was the 2019 version of the Green New Deal (GND-2019) as envisioned in the congressional resolutions?[7] The resolutions contain three major parts, beginning with a list of many critical environmental and economic trends. These include a reference to the latest reports on climate change, including an estimate that warming above 2°C will cause more than $500 billion annually in lost output by 2100. On socioeconomic factors, the resolution states that the United States is experiencing declining life expectancy, a four-decade trend in wage stagnation, and rising income and wealth inequality.

The second part lists the five major goals in the resolution. These are (1) achieve zero net greenhouse gas emissions, (2) create millions of high-wage jobs, (3) invest in infrastructure and industry, (4) secure several environmental objectives such as clean air and water, healthy food, and access to nature, and (5) promote justice and equity, particularly for frontline and vulnerable communities.

As outlined in the third part, the objectives should be implemented through a ten-year mobilization with a long list of goals and projects. Many of these are undefined and aspirational (economic security, healthy food, high-quality health care, family farming). Other goals are probably beyond the capability of new or old deals, such as "guaranteeing a job with a family-sustaining wage, adequate family and medical leave, paid vacations, and retirement security to all people of the United States."

Focusing on the Green part of the Green New Deal, we find several key proposals. First is the proposal to attain zero net global greenhouse gas emissions by 2050. This goal is based on current analyses of what would be necessary to attain the 2°C target for limiting global warming. It must be emphasized, however, that actually attaining zero net emissions is at the outer edge of the feasible, and the current agreed-upon policies in the Paris Accord of 2015 will not come close to that goal (as discussed in the chapters on Global Green).

A second concrete goal is "meeting 100 percent of the power demand in the United States through clean, renewable, and zero-emission energy sources." While no time line is specified, this might apply to either the 10-year window or the 2050 goal. To put this goal in context, it is useful to examine the current projections made by the federal Energy Information Administration (EIA) in its 2019 report.[8] Fossil fuels accounted for 61% of electricity generation in 2018 and are projected to fall to 55% by 2050. Even in the EIA's most optimistic case, fossil energy only falls to 41% by 2050. Current estimates suggest that moving to a zero net-carbon electricity system would raise the cost of power generation by 200 to 400% given current or near-term technologies.[9]

A third theme is the proposal for climate change: "Eliminating pollution and greenhouse gas emissions as much as technologically feasible." This goal of taking the maximum steps to eliminate emissions as is technologically feasible applies to infrastructure, manufacturing, agriculture, and transportation. What exactly does this mean? This approach does not apply a cost-benefit test to policies. Rather, it is one in which policies would take the maximum steps *at whatever the cost*. The language is reminiscent of early approaches to environmental legislation, which often used "best available control technology" or "maximum feasible" standards. Such approaches have been extremely challenging for regulators because they can, in principle, allow measures with extreme costs and minimal benefits.

This leads to a fourth point—what is excluded rather than what is included. From the vantage of 2020, the striking omissions include any discussion of using market approaches such as prices, taxes, or tradable permits as instruments of environmental policy. The GND-2019 resolution makes no mention of taxes or prices (carbon or other). It suggests that goals will be mandated by regulatory means, much like the first command-and-control environmental regulations of the 1970s. However, in recent times, rigid regulatory mandates have been increasingly displaced by market-type or market-supplemented regulations.

A critical omission is a discussion of the need for international coordination of policy, which is central for global public goods like climate change. Little progress on global issues can be made by domestic actions in the United States without building international coalitions. And it is striking that no mention is made of the only current successful regime for dealing with emissions, which is the European Trading System (ETS) for limiting carbon dioxide (CO_2) emissions.

Finally, the GND is primarily about policies to enhance equality and fairness rather than Green policies. Our analysis of Green policies emphasizes the centrality of market failures such as pollution and congestion. In other words, Green problems are centrally those that involve misallocation of public goods or failure to deal effectively with important externalities. These are distinguished from—although often related to—issues of inequality, unemployment, and inadequacy of private goods like housing and food. While some aspirations of the GND are directly aimed at Green goals, such as curbing global warming, most involve others such as reducing inequality and improved provision of private goods.

Thus, the GND describes a broad portfolio of policies—some targeted at promoting a Green society while most aim at the broader set of issues dealt with during the New Deal period.

Notwithstanding any criticisms or reservations, the 2019 version of the Green New Deal was a major political event and success for its sponsors. It highlighted the goals, particularly of climate-change policy, along with the need for policies to cushion low-income or heavily affected groups from adverse impacts. Unfortunately, it avoided the inconvenient truth that climate-change policies—particularly those that would meet its goals—would require aggressive price-raising measures, probably through carbon taxes. Confronting that truth would wait for another day.

PART IV

Green Across the Social and Economic Landscape

16

Profits in a Green Economy

The following chapters examine how Green can apply to many areas of the economy, the environment, and the broader society. These show the powerful potential of Green taxes and examine the importance of innovation in reaching our societal goals. We then examine ethical behavior for individuals, corporations, and investors.

One theme running through the discussion is the distortions that lead to Brown behavior—which generates pollution, congestion, and global warming. We also will examine the trade-offs between individual economic status or profits and societal welfare. Many of the distortions, particularly those that involve corporations and financial markets, arise because profits give misleading signals. A firm that maximizes profits will sometimes be led to production decisions that have harmful spillover effects. The root cause is often not the evil intent of the firm but the misleading compass provided by prices and profits.

Before considering the issues of taxes, innovation, and ethical behavior in the following chapters, we turn therefore to consider the role of profits in a private market economy. While this topic may seem remote from this book, it is central. Profits are the primary driving force behind market activity—both for better and for worse. Perhaps the biggest villain identified in this book is a distorted profit motive. With this thought in mind, let us examine the definition, measurement, and potential distortions of profits.

Different Views of Profits

Profits get a mixed reception in the public space. They are often seen as the result of unscrupulous thievery from customers or workers. Or from price gouging by firms in time of scarcity. Or from three-card-monte schemes that enrich management. In the electronic age, Facebook and Google have become enormously profitable by selling people's personal data without their knowledge and often without express permission or for allowing Russia to interfere in elections. Most important, corporations are charged with putting private profits above the public interest.

Pope Francis gave a theological critique of profits when he wrote, "Where profits alone count, there can be no thinking about the rhythms of nature, its phases of decay and regeneration, or the complexity of ecosystems which may be gravely upset by human intervention."[1]

Yet another view comes from the supporters of "free-enterprise capitalism." They often view profits as the reward for innovation and entrepreneurship. Here is the way the Chicago-school economist Milton Friedman explained it:

> After taxes, corporate profits are something like 6 percent of the national income. . . . The small margin of profit provides the incentive for investment in factories and machines, and for developing new products and methods. This investment, these innovations, provided the wherewithal for higher and higher wages.[2]

Publisher Steve Forbes put it more simply: "You don't get economic growth without investment. Capital comes from savings and profits. Period."[3]

The Economics of Profits

All these critiques and defenses contain some elements of truth. But they do not get to the economic function of profits in a market economy. Focusing on the production of private goods, we emphasized the invisible-hand principle of effective markets as one of the four pillars of a well-managed society. The invisible-hand principle states that in an ideal market economy, society is best off if firms engage in behavior that maximizes their profits. (Recall that the ideal is one in which the production of private goods operates without externalities and leaving aside inequality.)

Here is the reason. Begin with the textbook definition of profits: *Profits are the difference between the dollar revenues and dollar costs of producing goods and services.* This means that profits are the difference between the value of sales to consumers and the costs of production to workers and other producers. In increasing its profits, a firm is increasing the value of sales and reducing its costs. Doing so in an ideal market squeezes the most consumer satisfaction out of society's scarce resources. A crucial detail is that in the "ideal" market economy, dollar revenues and dollar costs accurately reflect social value.

When it applies, the invisible-hand principle greatly simplifies the ethics of individual and corporate behavior. It suggests that firms and individuals can operate without worrying that they are routinely hurting other people. All that is needed for socially responsible behavior in an ideal market is to behave as responsible members of the market community: to work hard and play by the rules.

However, before we give three cheers to the market, before we become too enamored of the invisible-hand principle, we need to remember its shortcomings. There are indeed externalities that distort the economy and might even have lethal effects. And the incomes generated by markets are sometimes extremely unequal and unfair. Therefore, let us give one cheer, but not three, for the market and for the role of profits.

Trends in Profits

Let us examine the trends in profits in the U.S. economy. For this purpose, we look at the nonfinancial corporate sector, which is the heart of the economy. It encompasses manufacturing, mining, communications, information, retail and wholesale trade, transportation, and much of services. These industries represent a little more

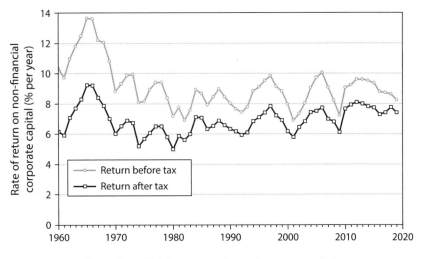

FIGURE 16-1. Trend in profits on U.S. domestic nonfinancial corporate capital

than half of the business sectors of the economy and about two-fifths of the entire economy.

For 2017, nonfinancial corporations (NFCs) owned $18.7 trillion of domestic capital (factories, machines, software). They earned $1.638 trillion of domestic profits and related income before tax and $1.383 trillion after tax. The rate of return on domestic non-financial capital was 7.4%. Figure 16-1 shows the series over this period. Table 16-1 shows the trend of the rate of return over the past half century along with the rate of taxation and the real return on safe Treasury bonds.

Three points stand out in the figure and table. First, the post-tax return on capital has been volatile but largely unchanged over the entire period from 1960 to 2019. While corporations are highly profitable, only modest changes have occurred in their overall profitability since the government began recording the data.

What about the increasing profits of the tech oligarchs? In reality, they comprise only a small fraction of corporate profits. The profits of all the giant tech companies such as Amazon, Facebook, Alphabet (Google), and Microsoft comprise only 12% of corporate profits— even though they make up a larger share of the stock market.

A second interesting trend is that the tax rate on corporate earnings dropped sharply over the period, from 31% of profits in the first

TABLE 16-1. Profitability of U.S. nonfinancial corporations

	Before-tax rate of return	After-tax rate of return	Average tax rate	Rate of return Treasury bonds
1960–1985	9.8%	6.8%	31%	2.2%
1986–2019	8.7%	7.0%	19%	2.4%

Note: Here and elsewhere, profits are defined in a broad sense to include all earnings on capital, including interest.

half to 19% in the second half, and actually hit a low of 10% in 2019. This downward trend masks the changes in the before-tax return, which declined substantially over the six decades.

The final point is that corporate capital is highly profitable relative to the real or inflation-corrected interest rates on safe bonds. Capital returns over the 1960–2019 period were more than 4 percentage points higher for corporations than for government bonds. It is generally believed that this difference reflects a premium required to compensate for the riskiness of corporate capital (the riskiness stemming from the uncertainty of corporate profits along with the volatility of the stock market).

However, for most people, the major surprise is that the return on corporate capital, while volatile, has shown no major trend over the last six decades.[4]

Dynamic Profits as Shareholder Value

Profits are a static concept, referring to the net proceeds during a single period, such as a year. But people are generally interested in the evolution of profits over time, or total current and future profits properly discounted. For a corporation, this is called *shareholder value*, which is the focus of corporate decisions and of corporate social responsibility.

Shareholder value is in fact quite simple. It is the value of all the shares of equity or common stock of the company and is sometimes called *market capitalization* or *market cap*. For example, in the fall of 2020, the shareholder value of Apple stock was $2,215 billion.

In the simplest financial theory, shareholder value is determined by expected future cash flows to shareholders, or more precisely, the present value of cash flows (dividends and stock buybacks). The present value takes the current and future cash flows and discounts them at the appropriate discount rate (discussed in chapter 13).

For readers who do not mind a digression to understand the mathematics of shareholder value, here is an example. Assume that ABC Inc. earns profits of $100 a year and distributes it all to shareholders. Further assume that the discount rate for ABC is 5% per year. Then the shareholder value (SV) and present value (PV) are $2,000.

$$SV = PV = \frac{\text{Profits}_1}{1.05} + \frac{\text{Profits}_2}{(1.05)^2} + \frac{\text{Profits}_3}{(1.05)^3} + \cdots$$

$$= \frac{\$100}{1.05} + \frac{\$100}{(1.05)^2} + \frac{\$100}{(1.05)^3} + \cdots = \frac{\$100}{0.05} = \$2000$$

The complete version of the idea that firms should maximize profits is that they should maximize the present value of profits, which is shareholder value.

Finance specialists quickly point out that this idealized view of shareholder value rests on many shaky assumptions. For example, those who buy and sell stocks do not know the cash flows in the future. They do not know future monetary policy or tax policy, so they do not know the discount rates. Perhaps the reported cash flows are manipulated by companies to make management look good or to pump up stock prices. All these concerns have some validity. But the basic point is useful to remember—that shareholder value depends ultimately on the firms' current and future profits, even though it is seen through a glass darkly.

Profits as the Compass of a Market Economy

Profits are like a compass that points the managers of a firm in a certain direction. This can be visualized in figure 16-2. Let us say north or up is for socially desirable goods and services, while south points to destructive activities like producing toxic wastes. Thus, we want our compass to point true north and not mislead us. Guided by a

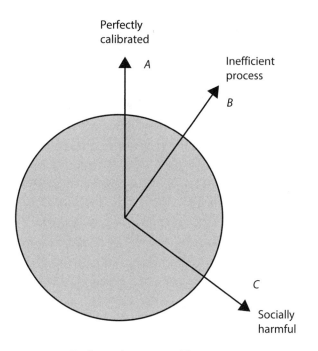

FIGURE 16-2. Profits are the compass of the economy

perfectly calibrated compass, the leader of the economic expedition will go in the right direction. In practice, the economy will go in a desirable direction if prices and incentives are properly calibrated.

However, suppose that the economic compass is damaged or mis-calibrated. This might arise because prices are misleading or because management incentives are distorted. The team may then head off in the wrong direction and produce inefficiently, as in point *B* in the figure. Or, in extreme cases, the compass might point in completely the wrong direction, as in point *C*, and lead to socially destructive products or processes. Thus, an accurate compass of profits is critical for guiding the economy in the right direction.

Recalibrating the Economy's Profit Compass

Why are profits like the economy's compass? To begin with, firms need profits to survive. Moreover, since the profits go to the firms' owners, they are interested in maximizing profits (or, in a dynamic version, maximizing shareholder value). Hence, since firms maximize profits, society needs to ensure that profits are a reliable measure of social value.

In a perfectly calibrated market, profits correctly measure net social value, which is the difference between social value and social cost. Social value is the amount that consumers are willing to pay, while social cost is the revenues that producers receive. Maximizing profits in a well-calibrated economy maximizes that net social value.

But suppose that profits are distorted because of missing elements. For example, perhaps the social cost of a good excludes the cost of pollution because pollution is unregulated and "free," whereas the true social cost of pollution is $Z per unit. Then, following the profit compass would produce too much of the good because its price was too low by $Z because of the excluded cost of the pollution.

An Example of Recalibrating the Profit Compass

The idea of recalibrating profits may seem esoteric, but it is fundamental to the effective operation of a Green economy. If profits provide the wrong signals, the economic locomotive goes in the wrong direction. This point can be seen using the example of electricity and carbon dioxide (CO_2) emissions, which are central to climate-change policy.

Table 16-2 illustrates the point. The first column shows the cost of operating a coal-fired generating plant in 2018 where CO_2 emissions are unpriced. Revenues were $60 per 1,000 kilowatt-hour (kWh; 1 megawatt-hour [MWh] = 1,000 kWh) and fuel and other variable costs were $32/MWh, so profits were $28/MWh. This calculation shows why coal plants continue to operate, even given the social costs, when the price of emissions is so low.

The second column shows how the profit calculus changes with carbon pricing. At a CO_2 emissions price of $100 per ton CO_2, this adds an additional cost of $81/MWh. Profits are then negative at and equal to minus $52/MWh. The plant would be retired at such a high penalty on emissions.

TABLE 16-2. Illustrative costs of the impact of externality pricing on the profits of coal-fired electricity generation

Costs, revenues, profits	Costs, revenues, profits ($/1000 kWh) at two different carbon prices	
	$0/tCO_2$	$100/tCO_2$
Revenues	60	60
Costs		
Capital, fuel, other	32	32
Cost of CO_2 emissions	0	81
Profits (revenues minus costs)	**28**	**−52**

Source: Estimates of costs are from the Energy Information Administration and are explained in detail in chapter 19.

Accurate estimates of the impact of externalities on profits is a complicated affair. One study led by Nicholas Muller used environmental and economic data to estimate the impact of the damages from air pollution on true prices and profits. In some cases the cost of pollution was so high that the true profits (or net social value of a product) were estimated to be negative. In looking at major industries, Muller et al. determined that the net social value of production was negative in seven industries because emissions regulations were so lax (implicitly, the price of emissions was set too low). These included not only coal-fired power plants (as in table 16-2) but also stone quarrying, solid waste incineration, sewage treatment plants, oil-fired power plants, marinas, and petroleum-coal product manufacturing.[5]

Overview of Correcting Profits

The point of correcting the prices and profit calculations is not primarily to penalize the companies. Rather, it is to give the proper signals so that firms will change their behavior. In the case of the coal-fired plant shown in table 16-2, the negative profit signal of high carbon prices will induce the managers to shut down the plant. It will also give signals for other low- or zero-carbon plants to start up, perhaps using natural gas or wind. It will further provide incentives for inventors and innovators to develop new and better low-carbon electricity technologies. All these examples reinforce the point about the critical nature of profits in pointing the economy in the right direction.

Another set of issues arises when a firm's managers have profit incentives that lead to poor decisions. Executive compensation is often based on the short-term performance of a firm, particularly the short-term increase in a firm's stock price. This incentive may lead to "short-termism" in decisions.

For example, a manager might defer investment projects that lower this year's profits even though they have a high long-term rate of return. In the worst case, in what chapter 20 calls the ninth circle of corporate irresponsibility, managers may defraud consumers with life-threatening products to keep profits up even though the truth will eventually come out and lower profits, or even destroy the firm.

Therefore, profits are like highway road signs showing the economy where to go. The purpose of Green management is to ensure that the road signs are accurate and do not lead the economy into dangerous territory. The next chapters apply these concepts in several areas: taxes, innovation, individual ethics, corporate responsibility, and ethical investment.

17

Green Taxes

Taxes have a poor image. George Washington, who became the first president riding on an antitax movement, argued that "no taxes can be devised which are not more or less inconvenient and unpleasant."[1] President Jimmy Carter said, "The federal income tax system is a disgrace to the human race."[2] President G.H.W. Bush declaimed, "Read my lips, no new taxes."[3] A tax on the large estates of the very richest Americans is labeled a "death tax." Presidential candidates have trillions of dollars of programs and subsidies but few tax dollars to pay for them.

Economists have a different view of taxes: they are the price we pay for public services. If you want a good public education for children, health care for all, environmental protection, and upgraded infrastructure, you need taxes to pay for these services. As Justice Oliver Wendell Holmes opined, "Taxes are the price of civilized society."[4]

People often think that taxes and public services live in different worlds. While that is true for individual items, it is wrong in the aggregate. The arithmetic is simple. In the long run, taxes must equal spending. More precisely, if a country does not default on its debt, the present value of taxes must be equal to the present value of spending.

This chapter has a simple message. Some taxes are less damaging and perhaps less painful than others—and indeed some taxes can be advantageous. Here is one way of putting this point. "There are some beneficial taxes. These would be taxes on bads, which can substitute for taxes on goods." This chapter will explain the logic behind Green taxes.

Tax Efficiency

Economists have been concerned with the efficiency of taxes for more than a century. The basic analysis is explained as follows in introductory economics. When a good or service is taxed, this raises the price to consumers and lowers the price to producers. This price shift will tend to reduce the level of output of that product. For example, it has been shown high cigarette taxes reduce smoking.

If the tax is on inputs, such as labor or capital, then it will lower the posttax earnings of those inputs and tend to reduce the supply. Conversely, companies will tend to move their operations to countries that with low taxes, so-called "tax havens," such as Ireland. The net effect of taxes or subsidies, therefore, is to distort the level of inputs and outputs away from taxed activities and toward untaxed activities.

Taxes are not uniformly distortionary, however. Taxes on capital, particularly in a world of open borders and mobile investment, tend to be the most distortionary. As an example, suppose that corporate capital is taxed at 50% of net income, while noncorporate capital is untaxed. The results will be a reduction in the amount of corporate capital until its pretax return doubles relative to noncorporate capital. With high corporate taxes, investment in real estate (lightly taxed because of special provisions) would increase while investment in manufacturing (heavily taxed because it has few tax breaks) would decline. The economy would have too many houses and too few factories.

Taxes on labor income are less distortionary. Studies have found that people tend to maintain their work hours when taxes reduce their posttax wages. Unlike capital, people tend to stay put. People are unlikely to emigrate from the United States to Ireland in response to higher taxes, so the distortionary effects of wage taxes are smaller than those on capital income.

Even less distortionary are taxes on *rents*, which are the returns to land and similar items that are fixed in supply. Because land is completely immobile, it will work for whatever it can earn. This means that land taxes have no effect upon land supplied, and there are no distortions at all from taxes on land rents. This interesting theory has been applied to the earnings of highly paid individuals (such as baseball players and business tycoons). Such highly paid people will work just as hard if their posttax earnings rise (as they have in the last two decades) or fall (should there be a wealth tax on billionaires).

Green Taxes

Where do environmental taxes fit into this spectrum from most distorting (such as on capital) to least distorting (such as land taxes)? Actually, they are off the spectrum. The reason is that environmental taxes reduce activities that society wants to reduce. Hence, a high environmental tax on, say, sulfur dioxide (SO_2) emissions will reduce the "production" of those emissions, which will reduce their damages. This means that Green taxes are beneficial—in other words, they increase economic efficiency—in contrast to virtually all other taxes, which reduce economic efficiency.

If they are beneficial, what is the appropriate level of Green taxes? Should they be set at the level that will maximize government revenues? Or for a fixed percent of needed revenues? Here is where the theory of optimal pollution comes into play. Our discussion of optimal pollution abatement entailed setting the price on pollution equal to its marginal damage. In the case of Green taxes, this implies that the most efficient outcome is that firms pay a tax on their pollution equal to the amount of external damage it causes. Hence, for example, suppose that public-health specialists have determined that the social cost of SO_2 emissions is \$3,000 per ton. Then, as a starting point, the efficient tax on SO_2 would be \$3,000 per ton.

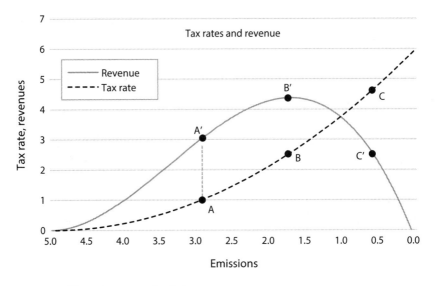

FIGURE 17-1. Taxes, damages, and emissions
Tax rates (*dashed line*) yield emissions on the horizontal axis and revenues on the vertical axis. The hill-shaped revenue curve shows that revenues have a maximum of *B'* of a little above 4 at a tax rate of about $2.5 and then decline at higher tax rates.

This leads to the central point about Green taxes. When the tax rate is set at the marginal damage of the pollutant, the tax will produce the first-best allocation of resources among goods, services, and abatement. It will internalize the externalities. Green taxes do not cause distortions. Rather, they *reduce* distortions because they reduce inefficient pollution. Suppose that an appropriate tax on sulfur emissions reduces the output or even closes a dirty coal plant. That change reflects the external costs that the sulfur emissions were imposing on the community, reduces the distortions from pollution, and improves overall welfare.

Figure 17-1 illustrates the basic analysis. It shows the tax rate and revenues on the vertical axis along with emissions on the horizontal axis. Suppose the government imposes a pollution tax of T on pollutant XO_2 equal to the marginal damage of pollution. Consider the case where the marginal damage and tax rate are equal to $1 per ton, shown as point A in figure 17-1. The revenues are the tax rate (T) times the posttax pollution, shown as point A' in figure 17-1.

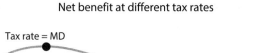

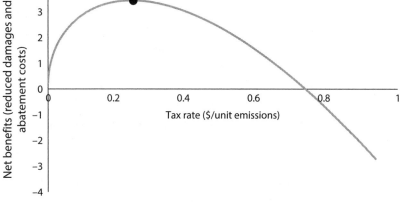

FIGURE 17-2. Net benefits are positive for Green taxes
Note that benefits rise up to the optimal tax rate for Green taxes.

Suppose that damages are higher, at point *C* in the figure. This would lead to revenues at *C'*. The surprising result is that the revenues are lower than at the lower tax and damage rate at *A*. The dome-shaped revenue curve shown as *A'B'C'* is the pollution Laffer curve (named for the economist Arthur Laffer).

For conventional taxes, raising tax rates so high as to reduce revenues—going beyond the maximum revenue point at *B'*—would be fiscal folly because it would both be highly distortionary and reduce revenues. In the case of Green taxes, the optimal tax might well be higher than the revenue-maximizing point. Consider the goal of many environmentalists that carbon emissions be reduced to zero. (Recall the goals of the Green New Deal, described in an earlier chapter, of zero net greenhouse emissions.) Perhaps a $500 carbon tax would achieve this. At a tax rate of $500, revenues would be zero. So we see a case in which the optimal Green tax would have zero revenues.

Figure 17-2 shows the net social value or benefit as a function of the tax rate. The highest net benefit comes at the point where the tax equals the marginal benefit, at the top of the benefit curve. For Green taxes, the top of the benefit curve has a positive altitude (indicating net gain from taxes), while for normal taxes the curve is always underwater, indicating net distortions.

Congestion Pricing: Theory and Practice

One of the most interesting applications of Green taxes has been in the development of congestion pricing, the darling of economists for years. It was originated by Columbia University's William Vickrey (1914–1996). He laid out the principles in a 1952 proposal for the New York City subway and won a Nobel Prize in part for his contribution.

A key concept here is that of congestion externalities. Let us look at an example to illustrate it. When roads are empty, the first car that enters will slow nobody down, so the cost of the travel of the first car imposed on others (the external cost) is zero. However, as traffic increases, every additional car will slow down the cars that follow. Suppose that upon entering the road I increase the time spent in traffic for the 120 cars behind me by one minute. If the value of people's time is $10 per hour, then I have imposed external costs of $120 \times (1/60) \times \$10 = \$20$. The more the cars and the longer they wait, the higher the imposed external costs.

Vickrey's basic idea was that people should pay for public resources—utilities, roads, airports, and other sectors—just as they pay for private goods like food, shelter, and entertainment. Moreover, the prices should reflect their external costs—the costs imposed on others. They should, according to Vickrey, vary over time depending upon the level of congestion, and they should be charged to all without exception. He envisioned what were at the time fantastic technologies for collection—those that have become commonplace with electronic tolling.

Vickrey admitted that his ideas were not well received by those who set public policy. "People see it as a tax increase, which I think is a gut reaction. When motorists' time is considered, it's really a savings." He insisted that the idea was not to reduce traffic but to increase it by spreading it more evenly over time.

Today, congestion pricing is mainly used in large urban areas such as Singapore, Milan, London, and New York. Most systems are extremely primitive and do not follow the Vickrey approach:

they are just "cordon tolls," where vehicles pay a charge for entering a city. For example, London has an £11.50 ($13) daily charge for driving a vehicle within the charging zone between 07:00 and 18:00, Monday to Friday. New York has a similar system. Singapore has moved closest to the Vickrey model, installing an advanced system with hundreds of electronic toll booths and tolls that vary by type of vehicle, time of day, and real-time congestion.

People often complain that congestion pricing has no impact. However, careful studies show that in fact pricing not only decreases traffic in peak hours but also increases traffic speeds. Perhaps most important for public support is that the revenues have been used to increase public transport, which further reduces traffic and pollution.

Congestion pricing as proposed by Vickrey was decades ahead of its time. Like other ideas that can solve key externalities, it may wait for many a year to gain the approval of Green elites and the public. However, as more cities and governmental agencies adopt it, and as populations become more comfortable and see the beneficial results, it has delivered the double benefits of reducing wasted time and energy and raising revenues for important public services.

The Potential for Green Taxes

The literature on environmental taxes is underpopulated relative to the vast literature on conventional taxes. What are the major potential sources of Green taxes? A review of work in this area suggests many areas where the externalities are underpriced. However, estimating the appropriate prices has proven extremely difficult, so we have only rough estimates.

The major fruitful areas are those where the externality (e.g., pollution) is well measured, where there is a convenient place in the production process to levy the tax, and where the administrative costs are small relative to the revenues. Particularly useful are greenhouse-gas emissions (especially carbon dioxide [CO_2]) and fuels such as gasoline, air pollution, and scarce public water. Here are the leading suspects, focusing on the United States, which has ample data and much environmental damage. Other taxes have either relatively modest tax bases or are much more difficult to implement.

CARBON TAXES

Among potential environmental taxes, by far the most important are carbon taxes. These have a large *tax base*, which refers to the value of the activity on which the tax is based. The carbon tax base is huge because annual CO_2 emissions in the United States are huge. Carbon taxes are attractive as a policy to slow climate change, as will be explained further in the treatment of climate-change policy, but they are also the preeminent environmental tax.

For the United States, here are the approximate estimates of the yield on a carbon tax. Emissions in 2019 were about 5 billion tons of CO_2 for industrial uses and another 1 billion tons of CO_2-equivalent for other gases, such as methane. We can use the U.S. government estimate of the marginal damage, which is $40 per ton. If emissions are unchanged, then total revenues would be 6 billion tons × $40/ton = $240 billion. However, emissions are likely to decline; at this price, emissions would decline by about 25% to 4.5 billion tons per year. This would produce about $180 billion per year, slightly under 1% of gross domestic product (GDP) or 8% of federal revenues at 2019 levels of economic activity.

If policy-makers want to increase the tax over time, say to $100 per ton, that level would increase revenues to about $400 billion per year. Peak revenues would be about $500 billion. Hence, carbon taxes could produce a substantial revenue stream—at least until they become so stringent that they shut down virtually all emissions and revenues.

We close with a reminder that carbon taxes are just a gleam in the eyes of environmentalists and fiscal specialists. The actual revenues in the United States and most other countries today is exactly zero.

Sulfur Dioxide and Other Air Pollution

Another potential source of revenue would be different conventional air pollutants. These would include not only SO_2 but oxides of nitrogen, carbon monoxide, and particulate emissions.

The United States currently gives away pollution permits, but fiscal specialists suggest they should be sold through auctions since these are valuable public property like oil or timber. The potential revenues from a SO_2 auction or tax can be estimated by examining emissions and trading prices for SO_2. We can calculate the implicit revenues as the revenues if the emissions permits were auctioned, which would be equal to the emissions times the trading prices. From 1994 to 2007, the average implicit revenues were almost $5 billion per year. After that, the price fell sharply because actual emissions were far below the regulated limits, so the implicit revenues collapsed.

These numbers are lower than the ideal, however, because the trading prices were much lower than the estimates of the marginal damage. The marginal damages were estimated to be around $3,000 per ton of SO_2, whereas the average price in the 1994–2007 period was $300 per ton. It seems likely that emissions would have declined sharply sooner with such high sulfur prices. At current emissions and the $3,000 price, the yield would be close to $10 billion annually.

The lesson here is that there are substantial potential revenues from SO_2 taxes, but they are far below those gained by a carbon tax.

The data on other pollutants are more difficult to ascertain because they are more sparse. Nitrogen oxides also have a trading program, and the implicit revenues here were in the range of $1 billion per year in the 2005–2010 period, after which there was a sharp decline in prices. The costliest pollution controls have been regulation of tailpipe emissions in automobiles, where the compliance costs were around $26 billion per year in 2010. If the regulatory approaches had been replaced by emissions taxes, these might have raised revenues in the tens of billions per year, but estimates here are inexact.

TRANSPORTATION EXTERNALITIES AND THE GASOLINE TAX

Automobiles are much despised by environmentalists. According to one study, externalities from automobiles include health damages, traffic congestion, accidents, air pollution, noise, climate change, habitat fragmentation, visual intrusion, degradation of nature and landscape, water pollution, soil pollution, energy dependency, and obesity.[5] While it might be possible to put Green taxes on each of these separately, it is probably convenient to tax an activity that has such a large army of harmful side effects.

The best approach would probably be to tax passenger-miles. This is difficult and intrusive, so most countries focus on taxing fuels—gasoline and diesel. These are plausibly related to CO_2 emissions, but the linkages to the other spillovers are more tenuous. Studies have found the total external effects to be between $1 and $4 per gallon, which is well above the U.S. fuel taxes and close to the taxes in Europe.

There are huge revenue possibilities in motor-fuel taxes. At present, the average tax rate in the United States is about $0.50 per gallon, which yields about $80 billion per year in taxes on motor fuels. If this were increased to $3 per gallon, this would yield about $370 billion a year.

So, as with the carbon tax, there is gold in the gasoline hills. Unlike a carbon tax, however, the gasoline tax is not an ideal Green tax. It would reduce environmental problems closely related to petroleum consumption (such as air pollution). But other issues (such as congestion or obesity) are unlikely to be effectively targeted by higher pump prices.

AUCTIONS OF SCARCE PUBLIC RESOURCES

Multiple other potential areas could be favorably affected by Green taxes (or more generally by resource pricing). The most obvious—and one I would implement virtually overnight—is congestion pricing in airports. If you have ever flown through busy airports such as Kennedy, O'Hare, or LAX, you have endured the long lines of airplanes waiting to take off. "Hi folks, this is your captain. We are number 34 to take off, so we'll wait at the gate and let you roast for 45 minutes. I'll keep you updated."

This malady is easy to cure. Just auction off the 60 slots for departures between 5:00 and 6:00 p.m. at O'Hare airport. The small planes or uneconomical flights to Milwaukee would choose not to fly, while the jumbo jet to London could easily absorb the charge. You can take the train to Milwaukee with little time penalty, but alternative modes from Chicago to London are hard to find.

Suppose airports raised $1 billion a year. The net effect would be less time on the ground and the ability for airports to modernize their facilities. Call this the *infrastructure relief fee*.

Other areas could benefit from the pricing of environmental resources. One is scarce water in the western United States. Water is modern gold there, and the nation is virtually giving it away to irrigate low-value agricultural products. If scarce public water were auctioned off to the highest bidder, the most valuable uses would have the necessary water while the low-value ones would find other uses for their land.

More generally, an interesting observation occurs if we look across the landscape. *Public resources are virtually all underpriced.* This includes not only the air, water, climate, subsoil minerals, grazing rights, and public lands but also less obvious items like landing slots, public highways, parks, and water. Applying the principles of Green taxes here will upgrade their use and raise revenues.

However, as a realistic second point, the fiscal yield on these public resources is likely to be modest, and the opposition is sure to be fierce. Securing the ability to price public resources will be house-to-house combat with the antitax groups as well as those who are short-sighted or want to keep their "free" public resources for themselves.

SIN TAXES

A final area that is important but not really environmental involves "sin taxes"—taxes on harmful products like tobacco, firearms, gambling, and alcohol. While these do involve some externalities (such as secondhand smoke, murder, financial ruin, and road accidents), the primary societal rationale on these is to discourage self-destructive behavior.

Sin taxes are currently substantial for tobacco, less so for alcohol, and virtually nonexistent for firearms and gambling. Taxes at rates

TABLE 17-1. Estimates of current and potential Green taxes for the United States

Externality	Current revenues	Potential revenues
	[Billions of 2018 $ per year]	
Climate change		
CO$_2$	0	159
Other GHGs	0	36
Ozone depletion	~0	1
SO$_2$	0	10
NOx	0	5
Other air pollutants	0	na
Water	0	[20]
Congestion	0	[20]
Motor Fuels	80	370
Tobacco	31	60
Alcohol	16	50
Firearms	2	40
Gambling	14	70
TOTAL	144	801
Total as % of federal expenditures	**4%**	**24%**

Numbers in brackets are estimates based on estimates of costs.

Source: Figures on the yield on existing Green taxes are generally from the Bureau of Economic Analysis and the Department of the Treasury.

Note: The numbers in brackets are rough estimates because reliable estimates are not available.

of 50% for both discouragement and to reflect social costs would bring in substantial additional revenues here.

SUMMARY ON THE POTENTIAL FOR GREEN TAXES

Table 17-1 shows a rough estimate of existing Green taxes along with their potential. Currently, Green taxes amount to $144 billion, or about 4% of federal revenues. The main areas for expansion are carbon taxes, fuel taxes, and sin taxes. At plausible rates to reflect the social costs, Green taxes could raise close to one-quarter of current federal revenues.

TABLE 17-2. Environmental taxes by category

Sector	Share of Green taxes in OECD, 1995 (%)
Transportation fuels	64
Motor vehicles	26
Heating fuels	5
Electricity	3
Waste	1
Other	1

The figure shows the major areas of Green taxes in OECD countries.

Source: OECD, *Environmentally Related Taxes in OECD Countries,* Paris, 2001.

Green taxes have the potential for a substantial amount of revenue if they are implemented in key areas. These taxes not only help pay for necessary government activities but also have the advantage of improving the functioning of the economy and society. Perhaps more important is that they can help achieve society's Green objectives (such as cleaning the air or slowing climate change) while minimizing bureaucratic regulatory approaches.

Green Taxes in Practice

If we look at Green taxes in practice, they are a hodgepodge of taxes of convenience in different sectors. Table 17-2 shows the averages for different countries along with the major sectors.[6]

A few points stand out. As table 17-2 shows, most environmental taxes are levied on road transport, either motor fuels or vehicles. These comprise about 90% of environmental taxes in all advanced countries, and an even larger share in the United States. But a further look shows that environmental taxes are a small part of revenues: environmental taxes comprise only 5% of all taxes for advanced countries.

Third, most environmental taxes are not pure environmental charges since they do not directly tax the externality. For example, a gasoline tax does reduce gasoline consumption, but it does not directly tax many of the externalities associated with transportation.

So where do we stand? How are governments using these taxes? Here are the basic results.

ON CARBON TAXES

The chapters on climate change later in this book will suggest that the marginal damage and optimal carbon tax is in the neighborhood of $40 per ton of CO_2. The World Bank estimates that the average carbon tax (or price) in major countries is about $2 per ton.[7] This includes both explicit taxes and the market price from carbon-trading regimes. The United States is not using a carbon tax at all, with its zero tax rate. Hence, this tax is essentially unused.

FOR SULFUR DIOXIDE

The United States and some other regions use a cap-and-trade system to limit SO_2 emissions. While the trading prices in the early years (after 1990) were substantial, in recent years they have fallen sharply. The actual price is far below the estimated marginal damage. Finally, since the allowances are given away to firms rather than auctioned, no revenues are collected. Hence, the major pollutant of SO_2 is not subject to Green taxation.

OZONE-DEPLETING CHEMICALS

One of the few true Green taxes is the U.S. tax on ozone-depleting chemicals like chlorofluorocarbons. The tax is proportional to the ozone-depleting potential of the product. While these have a genuine Green design, the tax rates are much lower than the marginal social cost.

Impact on Inequality

A standard concern with Green taxes is that they are regressive—that is, they have a larger impact on relatively poor households. The reason for the regressivity is that low-income households spend a larger fraction of their incomes on energy and other environmentally sensitive goods and services. The distributional impact of Green policies more generally was addressed for pollution control in chapter 4.

While Green taxes tend to be regressive, fiscal specialists offer a simple remedy. The revenues can be partially rebated to households in a manner that offsets the regressivity. An outstanding study by Gilbert Metcalf investigated possible combinations of Green taxes and rebates to determine a package that would be neutral across income groups. He found that a package of Green taxes would have a negligible impact on the income distribution if the funds are rebated to households through reductions in the payroll tax and personal income tax.[8]

Conclusion on Green Taxation

Green taxes are one of the clearest and cleanest examples of how Green thinking can improve the health and prosperity of countries. Green tax reform allows countries to pursue the twin objectives of raising revenues efficiently while improving the environment.

However, countries have seldom realized the promise of Green taxation and have largely ignored this powerful new set of taxes. Aside from gasoline taxation (worthwhile, but only indirectly related to environmental objectives), there are essentially no Green taxes. The most useful single environmental tax is a carbon tax—a tax that would move toward a central environmental objective, is easy to measure and enforce, and has the potential for large revenues. Other examples—such as taxes on conventional air pollutants, congestion, water, and other resources—are useful but more complicated and have smaller revenue consequences.

The summary here is that Green taxes are one of the most promising innovations of recent years. They are the holy trinity of environmental policy: they pay for valuable public services, they meet our environmental objectives efficiently, and they are nondistortionary. Few policies can be endorsed with such enthusiasm.

18

The Double Externality
of Green Innovation

The hulls of ships are fertile places for the growth of plants and animals. This nuisance, called fouling, costs on the order of $3 billion annually. The main compounds traditionally used to control fouling were organotin antifoulants, such as tributyltin oxide (TBTO). While effective, they are highly persistent, accumulating in the environment and causing damage to shellfish.

Maritime authorities worked to ban organotin-based agents. In response to this regulatory threat, Rohm and Haas Company searched for an environmentally safe alternative to organotin compounds, settling on Sea-Nine, which degrades quickly and has essentially no bioaccumulation. The company was awarded the Presidential Green Chemistry Challenge Award by the U.S. Environmental Protection Agency (EPA) in 1996 for this new product.[1]

Sea-Nine was Green by design. The broader point is that meeting the environmental challenges of the future will require technological changes. Ship hulls are just one example of the complex processes and incentives that lead to environmental innovation.

Looking at the COVID pandemic as an example, the most critical innovation was a set of safe and effective vaccines. The benefits of securing an effective vaccine one year earlier are literally in the trillions of dollars. The developers of successful vaccines are likely to make mountains of money, perhaps a few billion, but in reality, they will reap only a tiny fraction of the social benefits. This gap between social and private return is a major impediment to effective innovation, and indeed the gap is, as we will see, larger for Green innovations than for those for the regular economy.

Another environmental innovation was for sulfur dioxide pollution from electricity-generating plants. As we saw in the chapter on Green politics, emissions in that sector in the United States declined sharply. This decline resulted from many factors: using cleaner coal, substituting natural gas for coal, cleaning the sulfur from stacks, and using economic incentives to close down the dirtiest plants, as well as conserving energy. Each of these built upon a technological or institutional innovation and was driven by regulation or high regulatory prices on sulfur emissions.

Reducing congestion can be achieved with new tools like congestion pricing and electronic tolling. If we look backward to the filth of an earlier age, recall that the introduction of the automobile was central to removing a mountain of horse manure from our city streets.

Perhaps the greatest challenge is the need to reduce greenhouse-gas emissions, perhaps moving to zero emissions in the next few decades as suggested in the Green New Deal. Attaining this goal will require dramatic changes in energy technologies.

In the longer run, then, we will look to technological change to play a central role in implementing the Spirit of Green. This chapter discusses the challenges, including the central issue of the double externality.[2]

Green Design of New Products

A few years ago, I was sitting in a Yale College faculty meeting, looking at the proposed new courses. On the list was "Green Chemistry." I had never heard of this subject. What was it?

A little reading found the answer as described by the founders:[3]

Green chemistry comprises two main components. First, it addresses the efficient utilization of resources and the concomitant minimization of waste. Second, it deals with the ecological, health, and safety issues associated with the manufacture, use, and disposal or re-use of chemical products. The underlying tenet is "benign by design," with emphasis on pollution prevention through waste minimization as opposed to the end-of-pipe solution, waste remediation.

The tenet of "benign by design" highlights the importance of innovation in promoting Green principles, such as designing new products that preserve function while reducing toxicity. We saw that in the example of Sea-Nine as a successful application of Green design.

However, we must emphasize the strong headwinds faced by Green innovation. Research, development, and innovation for environmental products and services have a special challenge in what can be called a *double externality*. The reason is not only that clean production is underpriced but also that the private returns to innovation are below the public returns.

Let us unpack the central point here. The first externality occurs when the social cost of a good or service differs from its private cost. Take the problem of air pollution. If you travel to the large cities of India or China, you are likely to experience dangerous air pollution. Public health specialists estimate that millions of people die prematurely because of air pollution in these two countries. The air pollution results from little-regulated emissions, generally from coal-fired electricity generation. Those who produce and consume electricity do not pay the costs of the health damages from this technology. In other words, the market price of burning coal is below its true social cost.

A similar externality comes in climate change. Virtually everything we do directly or indirectly involves the consumption of energy; when that energy comes from the combustion of fossil fuels, carbon dioxide (CO_2) is emitted into the atmosphere. Here again, those who benefit from the energy consumption do not pay all the current and future costs imposed by those emissions. Thus, underpriced pollution is the first externality, which is thoroughly discussed in other chapters and well understood among those who study environmental issues.

The second externality—involving the research, development, and design (RD&D) of Green goods and services—is more subtle and involves the nature of new knowledge. New designs and innovations are what are known as public goods, which involve *positive* externalities. A public good is one that meets two conditions: the cost of providing the good or service to an additional individual is near zero (*nonrivalry*), and it is impossible or expensive to exclude individuals from enjoying the good or service (*nonexcludability* or *inappropriability*). While these words are inelegant, they are central to the properties of new knowledge.

All new technologies have these key properties. They exhibit nonrivalry because the use of a new design by one firm does not prevent its use by another firm. And they exhibit nonexcludability in that once a technology has been developed and disclosed, other firms cannot easily be excluded from using it.

The central distinction between knowledge and conventional goods is nonrivalry. Conventional goods are rival since when I eat a slice of bread, no one else can eat it. But ideas are nonrival because they can be used simultaneously by any number of people. While bread is scarce, existing ideas are not. Ideas are not depleted by their use. Indeed, the repeated use of a new technology (such as vaccinations or smartphones) often makes the new technology easier to use and more valuable.[4]

An important example of nonrivalry is the concept of vaccination. The modern medical use of vaccines is credited to Edward Jenner, who used cowpox material to create immunity to smallpox. Once the idea of vaccination was invented and understood, it could be used again and again to save millions of lives. The most lethal pathogen of all history was smallpox, which has been eradicated by vaccination. As I write this in 2021, people around the world are anxiously awaiting the results of population vaccination for COVID-19. The different COVID-19 vaccines build on earlier discoveries, science, and successful and unsuccessful vaccines. These earlier ideas are available to all in the dramatic race for COVID-19 vaccines.

Additionally, ideas are ultimately nonexcludable just because they are nonrival. In reality, valuable new designs encounter practical and legal obstacles that slow immediate diffusion, so for a while, inventors can exclude others at least partially. But, over time, valuable ideas eventually leak out around the world.

An example of the attempt to prevent leakage of technologies was Britain's attempt to limit the export of machine technology. This was enforced by prohibiting the export of textile machinery and even forbidding textile workers from leaving the British Isles. The prohibitions were in force from the 1780s to 1824. The penalties were fines on the order of 10 times average annual wages and up to 10 years in jail. Even these harsh measures were ineffective. People left the country, machines were taken apart and smuggled out, and machinery plans were increasingly available abroad. As one study concluded:

> Britain's prohibitory laws thus failed signally to stem the flood of technological information spreading abroad, either via men or machines, in this early industrial period. Administering and policing the sort of protection envisaged by the laws required Draconian measures that public opinion would not tolerate and internal economic and social conditions could not support.[5]

Nonrivalry and nonexcludability are the reasons for the innovational externality. However, this is an example of a beneficial externality. Inventors cannot appropriate for themselves the full gains from new knowledge—they cannot force others to pay for the full value of its use. As a result of the inappropriability of the full value, the private returns to innovation are usually well below the social returns. As a result, less innovation takes place than is optimal for society as a whole.

Table 18-1 shows the results of an important study by Edwin Mansfield and coauthors examining the social and private returns for 17 innovations. The bottom line shows their estimate *that the social returns were more than two times the private returns*.

Other studies suggest that the gap is larger for important innovations than for small ones, for more fundamental inventions than for minor improvements, and for innovations that competitors can

TABLE 18-1. Social and private returns to innovation

	Rate of return (percent)	
Innovation	Social	Private
Primary metals innovation	17	18
Machine tool innovation	83	35
Component for control system	29	7
Construction material	96	9
Drilling material	54	16
Drafting innovation	92	47
Paper innovation	82	42
Thread innovation	307	27
Door control innovation	27	37
New electronic device	Negative	Negative
Chemical product innovation	71	9
Chemical process innovation	32	25
Chemical process innovation	13	4
Major chemical process innovation	56[a]	31
Household cleaning device	209	214
Stain remover	116	4
Dish washing liquid	45	46
Median	**56**	**25**

A study of key innovations found that innovators did not capture even half of the social returns.

Source: Edwin Mansfield, John Rapoport, Anthony Romeo, Samuel Wagner, and George Beardsley, "Social and Private Rates of Return from Industrial Innovations," *Quarterly Journal of Economics* 91, no. 2 (1977): 221–40.

easily imitate. This result—the existence of a large gap between the social and private return to inventive activity—has been replicated in dozens of studies and is a staple finding of economics.

The Double Externality of Innovation: Pictures in an Exhibition

The double externality can be illustrated to show how inappropriability and environmental externalities sharply reduce the profitability of Green innovations. The discussion here applies specifically to *market-oriented* innovation—that is, to knowledge generation in sectors that respond primarily to profits and market incentives.

We can use the numbers in table 18-1 to illustrate the issue. The innovations listed there had an average private rate of return of 25% per year. Let us assume for simplicity that this is the average private return on all investments of any kind since profit-maximizing firms will make investments that are equally profitable on the margin. (Here and below, the rates of return discussed here are always assumed to be adjusted for taxes, subsidies, excessive discount rates, risks, and uncertainty.)

However, the average social return was much higher, at 56% per year. The difference arose because the inventor lost profits when imitators took some of the markets or when prices fell, and consumers benefited from the innovation.

Let us next consider a Green innovation. Perhaps it would be a new turbine design that saves fuel. It pays a 25% private return, but some of the gains are lost to the innovator, so the social return is 50%. However, there are further gains because the greenhouse-gas emissions and other pollutants are reduced. These add further social gains but cannot be captured by the inventor because CO_2 and other emissions are not priced or are underpriced. Perhaps, if we add the value of the environmental improvements, the total social return is 100%. The double externality has raised the gap between social and private returns from 25% to 75%.

Figure 18-1 shows schematically how the appropriability of the gains from innovation-related activity varies with the type of activity. The horizontal axis represents the *innovation spectrum*, a qualitative variable that indicates where an activity lies along the range from pure research through applied research, development, and so on to production. The vertical axis represents the appropriability of the gains from each activity, or the ability of the firm conducting the activity to capture its full value.

Pure research has exceptionally low appropriability, both because it has few immediate benefits and because the results can generally not be patented. An example of an inappropriable case is the discovery of laws of nature, such as gravitation or DNA. At the other end of the spectrum is production, such as manufacturing shoes or

Appropriability

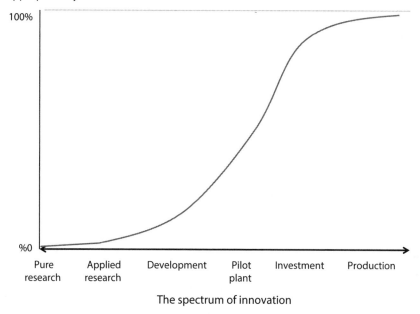

The spectrum of innovation

FIGURE 18-1. The spectrum of innovation and appropriability
Appropriability, or the ability of innovators to capture the full benefits of their efforts, is near
zero for pure research and close to 100% for production.

socks. This kind of activity has few externalities and is close to 100%
appropriable. Other kinds of innovational activities fall in between.

We can use a similar graph to depict the social and private returns
to these different kinds of activity. The horizontal line at the bottom
of figure 18-2 shows the private rate of return, which tends to be
equalized by market competition across different kinds of investments in knowledge and capital.

The downward-sloping curve at the top of figure 18-2 shows the
social return to investments in "normal" sectors, those whose products are not subject to environmental externalities. For these sectors, toward the right side of the figure, social returns are close to
the private returns because spillovers are low and appropriability
is high. At the other extreme, basic research has a large divergence
between social and private returns because appropriability is low.

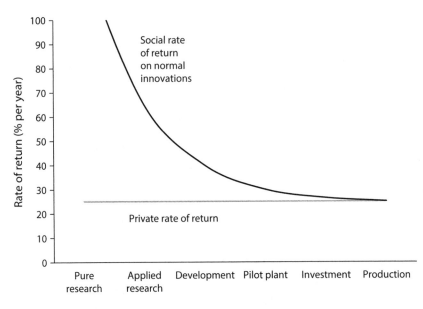

FIGURE 18-2. Social and private returns for different kinds of activities
The divergence between social and private returns varies according to the degree of appropriability. At the pure end of the spectrum, researchers gain an exceedingly small fraction of the market value of their discoveries.

Figure 18-3 shows how an environmental externality changes the incentives for Green innovations. For these, the private rate of return on innovations that are undertaken is again at the bottom horizontal line. For Green innovations, the social return is above the private return *and* above the social return for normal sectors. It is super high because the environmental externality adds to the knowledge externality, thus widening the gap between private and social returns.

Impact of Green Policies on Incentives for Green Innovation

One of the major surprises about Green policies is how they can affect the incentives for Green innovation. Recall that a central policy proposal to correct for spillovers is to "get the price right." This involves primarily setting prices on externalities so the market price equals the social cost of the activity. In the case of global warming,

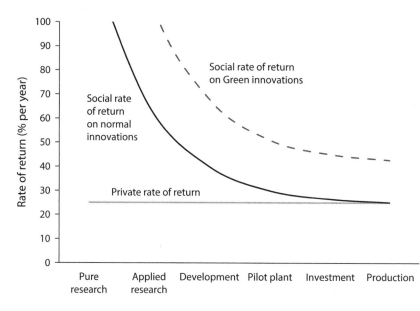

FIGURE 18-3. Social and private returns for normal and Green innovation
The divergence between social and private returns is much higher for Green innovations than for normal innovations because they suffer from the double externality of inappropriability and underpricing of the environmental gain.

the appropriate policy is to set the cost of greenhouse-gas emissions at the level of damage caused by emissions.

Assume that the government implements a policy of pricing the externality. As a result, when the externalities are corrected, there is no longer a double externality for Green innovation but only a single one—the knowledge externality that is experienced by innovators in all sectors.

The point is illustrated in figure 18-4. Assume that the environmental externality has been internalized by a governmental regulation. With the removal of the environmental externality, the dashed curve representing the social rate of return for Green innovations shifts to the left.

A gap will still remain between the social and the private returns on innovation in both the Green sector and the normal sector. However, the size of the gap is now determined by the size and nature of the innovative activity rather than by the sector. For example, there may be sizable spillovers from demonstration plants, or from basic

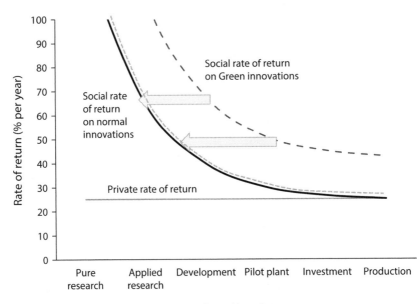

FIGURE 18-4. The social and private returns after public policies
When government actions internalize the externality by setting the price of pollution and other spillovers at their social cost, this corrective step removes the second externality from Green innovations. Green innovations share the same problem as normal innovations in that the returns to innovative activity cannot be completely captured or appropriated.

research into Green technology, but these spillovers would be the same as those in machinery or computers or other normal industries.

The central message of this chapter is that Green RD&D are doubly cursed by externalities, one for pollution and another for innovation. Appropriate pollution-correcting measures can cure only one of the two curses but leave the innovation externality untouched.

Innovation for a Low-Carbon Economy

To illustrate the issues involved in Green innovation, the balance of this chapter discusses the challenge of making a transition to a low-carbon economy. This is one of the central goals of climate policy today. While a full discussion is postponed to later chapters, we can carve out the innovation issues for this chapter.

- The first question is the challenge of decarbonizing the global economy. Is this just a routine question, like replacing typewriters with computers, or Edison light bulbs with new LED bulbs? Or is it more difficult and costly?

- A second question is the technological one. Today's economy is driven largely by fossil fuels like oil and coal. What technologies will replace these fossil workhorses of the modern economy? What fuel will propel our airplanes and heat our schools in the low-carbon world? What are the roles of nuclear, solar, wind, and other low-carbon fuels for electricity generation? These are exciting questions that engage engineers and scientists around the world.

- The third question from economics is subtle but equally important. How will we get firms to invent, develop, and commercialize these new technologies? How can we persuade consumers to buy and use these new technologies? It is not enough to have a bright idea about a solar-powered water heater or a carbon-eating tree. If firms are to invest billions of dollars in developing such technologies, they must find them profitable to produce and sell. Consumers must find them advantageous to buy. What are the mechanisms that will set in motion this chain of invention, production, and purchase of new low-carbon technologies?

The Challenge of a Low-Carbon Economy

Begin with the first question of the challenge of decarbonizing the economy. One of the goals of climate policy in many countries is to limit global warming to 2°C. Calculations indicate that this would require achieving zero net global emissions of CO_2 and other greenhouse gases by around 2050. The world is nowhere near achieving that goal. In fact, global CO_2 emissions have been growing, not declining in recent years. The world gets about 80% of its energy from fossil fuels today, much of that used in long-lived capital such as houses and power plants. How big is the challenge of reaching zero net emissions by 2050?

The short answer is that it is somewhere between highly unlikely and infeasible. We would need to replace a substantial part of the world's capital stock over the next three decades. The economic impact has been evaluated in several studies. An important study by the Energy Modeling Forum looked at the cost of the 2°C target in a range of models and different technological assumptions. The costs (present value of losses discounted to 2010) varied from $40 trillion to $500 trillion for the most and least optimistic technological assumptions, respectively.[6] Other studies indicate that achieving the 2°C target is infeasible without the unlikely combination of drastic changes in global policies and extremely rapid technological change.

Promising Technologies

Given the enormous scale of the transition needed to achieve a low-carbon economy, what are the promising low-carbon energy sources? This is a major area of research today by scientists and engineers, and we can only scratch the surface of the topic. However, a few remarks will illustrate the nature of the transition.[7]

A good place to start is with the current and prospective costs of different kinds of electricity generation for the United States. Table 18-2 shows estimates by the Energy Information Administration (EIA), which is the source of the best energy data for the United States.[8] This illustrates the costs of current and prospective technologies in dollars per 1,000 kWh. The three numerical columns contain the costs of generation for three different carbon prices (or carbon taxes). The first shows the costs with the current price in the United States and most countries of zero dollars per ton of CO_2 which represents no climate policy. The last two columns display the impact of low and high carbon prices. At the low end is the price recommended by the U.S. government ($40 per ton CO_2), while the high end is a price that would be consistent with aggressive emissions reductions ($200 per ton CO_2).

There are three categories:

- The top list is for existing power plants. For these, the capital costs are bygones, so the only costs are fuel and other current costs.

- The second block represents currently available technologies.
- The third list is for technologies under development.
  Some are under development (such as *advanced combined cycle*, described below), while others will require years of development and testing (such as advanced nuclear).
- The bottom line is the current average cost of electricity, at $41 per 1,000 kWh.

First consider the most economical current technologies with no climate policy (a carbon price of $0 per ton of CO_2). All four technologies shown in the table are economical at the current average cost of $41.

For new plants and current technology (the second block), the first three are reasonably economical, but conventional coal becomes uneconomical because of regulatory costs. The dominant technologies are natural gas (combined cycle) and onshore wind. Indeed, these are the most rapidly growing sources over the last decade.

Next, look at the last column, which shows the costs with a strong climate policy and a carbon price of $200 per ton CO_2. Currently, the only mature low-carbon technologies are renewable wind and solar power. Coal and natural gas have costs that are three to five times today's costs when the carbon price is included. However, the renewables have severe limits in terms of technical issues (such as load curves) as well as long-run supply limitations. Note as well that replacing the current electricity structure with renewables would be a massive undertaking because renewable electricity represents only a small part of total generation, about 10% of the total in 2018.

If we look at future technologies, two might enter the stage: combined-cycle natural gas with carbon capture and sequestration (CCS) and advanced nuclear power. These would cost about two times current costs, but in principle they would be able to scale up to economy-wide production levels. Moreover, they are still a long way from being ready for large-scale deployment. There is today not a single large-scale plant with advanced CC with CCS or advanced nuclear power, so introducing these on a large scale would realistically take decades.

TABLE 18-2. Estimates of the cost of electricity generation with alternative carbon prices

Plant type	System cost ($/1000 kWh)		
	At $0/tCO_2$	At $40/tCO_2$	At $200/tCO_2$
Existing			
Solar PV	12	12	12
Wind, onshore	16	16	16
Conventional coal	26	58	187
Conventional CC	37	51	105
Current			
Conventional CC	46	60	114
Wind, onshore	56	56	56
Solar PV	60	60	60
Conventional coal	75	107	236
Future			
Advanced CC	41	55	109
Advanced CC with CCS	68	69	75
Advanced nuclear	77	77	77
Coal with 30% CCS2	104	130	232
Coal with 90% CCS2	127	132	151
Current average cost	41	NA	NA

CC = Combined cycle (natural gas)

PV = photovoltaic

CCS = carbon capture and storage

NA = not applicable (because CO2 price = $0)

This table shows estimates of the costs of different kinds of power generation with different carbon prices. The first group shows current generating facilities. The second shows new facilities with currently available technologies. The third group shows estimates of technologies that may become available in the coming years.

Source: Estimates of levelized costs are from the U.S. Energy Information Administration; costs from carbon prices added by author.

Table 18-2 is worth careful study as it shows the major challenges that must be overcome to make the transition to a zero-carbon economy in just the electricity sector. The main conclusions are the following: First, energy in a zero-carbon future will cost significantly more than today's production. Second, the country will need to replace a substantial fraction of its electricity capital stock to reach zero emissions. And third, the best long-run solution will require developing new technologies that are expensive and will put major burdens on the regulatory and economic systems of countries.

But all these estimates must be viewed with caution. We cannot reliably see far into the future, and technologies are developing rapidly in many areas. Therefore, we need to be attuned to new possibilities. Even more important, we need to encourage fundamental and applied science and ensure that markets provide the appropriate incentives for inventors and investors to discover and introduce new low-carbon technologies. And that issue leads to the final section of this chapter, which explores governmental policies to promote innovation.

Promoting Low-Carbon Innovation

Most decisions on energy and the environment are made by private businesses and consumers on the basis of prices, profits, incomes, and habits. Governments influence decisions through regulations, subsidies, and taxes. But the central energy decisions are taken in the context of market supply and demand.

When we think of energy and environmental decisions, we usually think about a new car, new appliances, or renovating our houses and factories. All of these take place within *existing designs and technologies*. However, as the last section showed, over the longer run, moving to a Green economy includes critical decisions about *new and undeveloped technologies*. For example, rapid decarbonization will require substantial changes in our electricity-generation technologies, including profoundly different ones such as CCS.

How do technological changes arise? The answer is, usually, through a complex interaction of individual brilliance, persistence, economic incentives, corporate structure, and market demand.

Solar power, for example, typifies the meandering history of most fundamental inventions. The story begins in 1839, when the young French physicist Edmond Becquerel hit upon the photovoltaic effect while experimenting with an electrolytic cell. The physics underlying the photoelectric effect was explained by Albert Einstein in 1905, for which he won the Nobel Prize.

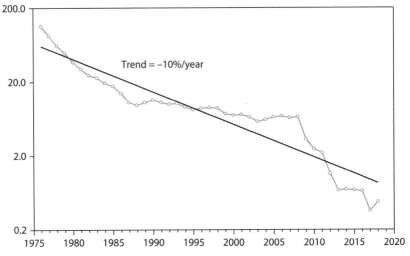

FIGURE 18-5. The price of solar power has declined sharply for almost half a century.
Source: National Renewable Energy Laboratory.

The first important practical applications for the photovoltaic cell waited for more than a century after Becquerel's discovery. Scientists at Bell Telephone Labs developed solar cells in the mid-1950s, and governments got involved as they realized the potential of solar power for use in space satellites and remote locations.

At that point, solar technology blossomed, with applications in space satellites, small arrays on houses, and large solar plants. Efficiency (energy of illumination per unit of sunlight energy) rose from 4% in the first solar cells to 47% in the best current applications as of 2020. Costs have fallen dramatically since the first cells. Figure 18-5 shows the trend in the price of photovoltaic modules, which have declined at 10% per year since 1976. As table 18-2 shows, solar PV power is competitive with the most economical fuels today at even modest carbon prices.

Let us return to the issue of the double externality in Green innovation. Investments in low-carbon technologies are depressed because the private returns to innovation are below the social returns, and private returns are further depressed because the market price of carbon is below its true social cost.

Our discussion of low-carbon technologies suggests that a low- or zero-carbon world will need new technologies like CCS. What exactly is CCS? The following description is based on a careful study by a team of engineers and economists from the Massachusetts Institute of Technology (MIT). The basic idea is simple. CCS would capture the CO_2 at the time of combustion and then transport and store it in some location where it would remain for hundreds of years and thus not enter the atmosphere.

We will use the example of coal because that is the most plentiful fossil fuel and a leading candidate for a large-scale CCS deployment. Engineers think that CCS with natural gas will be less expensive at today's natural gas prices in the United States, but the basic principles outlined for coal are similar for natural gas.

We can simplify by assuming coal is pure carbon. Then the basic process is expressed as the chemical reaction:

$$\text{Carbon} + \text{oxygen} \rightarrow \text{energy as heat} + CO_2$$

Hence, combustion produces the desired output (heat that can be used for electricity generation) plus an undesirable by-product, CO_2.

The trick is to capture the CO_2 molecules before they enter the atmosphere. CO_2 separation is currently in operation today in oil and natural gas fields. However, existing techniques operate at a small scale and are not yet ready for deployment in large coal-fired electrical generation.

One promising technology is integrated gasification combined cycle (IGCC) with CO_2 capture. This process would start with pulverized coal, gasify it to produce hydrogen and carbon monoxide, further react the carbon monoxide to produce highly concentrated CO_2 and hydrogen, separate out the CO_2 with a solvent, compress the CO_2, and finally ship the CO_2 and store it. All this sounds complicated, and it is, but it is not much more complicated than the technologies currently used in generating electricity from coal.

The major issues with CCS are cost and storage. The effect of CCS on the cost of electricity is shown in the last group of technologies in table 18-2. The cost of combined cycle when CCS is added rises by 63% (from $41 to $68 per 1,000 kWh).

While CO_2 capture is the expensive part of the process, transportation and storage are likely to be the more controversial parts. One problem is simply the scale of the materials that would be stored. The most prominent storage sites are porous underground rock formations such as depleted oil and natural gas fields. Another issue is the risk of leakage. This would not only reduce the value of the project (because the CO_2 would enter the atmosphere) but could pose problems for health and safety. My favorite option would use gravitational storage in the deep oceans. If the CO_2 is deposited in the deep ocean, the CO_2 would be heavier than water and would remain there for many centuries.

At present, CCS faces many hurdles. To make a substantial contribution, it would need to remove tens of billions of tons of CO_2 each year, yet, currently, only 25 million tons are removed annually. So, it would need to be scaled up by a factor of close to 1,000. Moreover, there are inadequate data on the performance of underground storage, and extensive experience is necessary to ensure scientific and public acceptability. People are frightened of the prospect of a huge burp of CO_2 causing unforeseen damage.

As for many other large-scale and capital-intensive technologies, CCS is caught in a vicious cycle. Firms will not invest in CCS on a large scale because of a vicious cycle of reinforcing factors. It is financially risky, public acceptance is low, it faces big regulatory hurdles to large-scale deployment, and there is so little experience with CCS at a large scale. Breaking out of this vicious cycle is a major dilemma for public policy in this as in other new, large-scale energy systems.

The critical point here concerns the impact of externality prices on the incentives to innovate. Assume for the moment that CO_2 can be removed at a cost of $100 per ton CO_2. If the price of CO_2 is zero, then the plant would lose money. No profit-oriented company would invest in this process if it knew the price of CO_2 would be zero forever.

However, suppose that a firm thought that countries were going to implement an ambitious global-warming policy—one in which the price of carbon would predictably rise to $200 per ton in a few years, as in the last column of table 18-2. At that price, businesses would estimate that operating a CCS plant would be profitable. The firm would be producing CO_2 at a cost of $100 a ton and in effect selling it to the government at a price of $200 a ton. Firms would proceed cautiously, looking at different approaches, but they would have economic reasons to invest in this technology. This same logic would apply to investments in solar, wind, geothermal, and nuclear power. Indeed, the same point applies more broadly to Green innovations.

———

This chapter leads to three major conclusions. The first point is that Green innovation suffers from a double externality. Not only are there inadequate returns to the production of Green goods and services (such as those that degrade quickly or lower greenhouse-gas emissions), but there are diminished incentives to undertake innovative activities to design new and improved Green processes and products because of the wide gap between social and private returns to research.

Second, many of the Green challenges faced today will require deep technological changes, whether scientific, engineering, or institutional. We saw this in reviewing potential technologies for a zero-carbon electricity sector, where the major large-scale technologies are not yet available on a large scale.

Third, progress toward meeting our Green goals depends upon the innovations of profit-oriented firms, and they in turn must have the proper incentives to make innovative activities profitable. This can best be accomplished by ensuring the internalization of major externalities, such as by putting a price on pollution. For example, carbon prices must be sufficiently high that investments in low-carbon technologies can expect tangible and secure financial payoffs. Without high carbon prices, innovators and firms will not be motivated to invest in low-carbon technologies. Thus, ensuring remedies for externalities has the further benefit of giving impetus to Greener new technologies in the future.

We can put these points in a broader context. The country may have the best climate scientists developing the most skillful projections of climate change; it might have the best materials scientists working on high-efficiency CO_2 pipelines; it may have the best financial wizards developing new financial derivatives to fund all these investments. But if the carbon price is zero, then projects to develop promising but costly low-carbon technologies will die before they get to the boardroom of a profit-oriented company.

Individual Ethics in a Green World

The Green movement has a large normative component. "You should minimize your carbon footprint. We should preserve important habitats and species. I believe in preserving the natural world for our grandchildren. We should be responsible homeowners and investors."

These norms underpin many issues that arise in this book, particularly those on corporate and investor responsibility. Given the ethical dimensions, we should at the outset stand back and ask, "What is the essence of Green ethics? What are its important assumptions and precepts? How can we apply it to different areas?" Addressing those questions is the purpose of this chapter.

Ethics is an enormous field, one whose roots go back to the Bible and Aristotle, with a long line of thinkers from the Catholic Church, as well as important philosophers of the Enlightenment and the modern era. At the most general level, ethics involves the systematic conception of right and wrong behavior. Ethics involves both general principles ("Do no harm") and applies to specific fields such as abortion, human rights, and war. We will limit this discussion to ethics as it applies primarily to economic, political, and environmental concerns and leave aside many of the other weighty issues.[1]

Ethical Federalism

Some writings on ethics deal with the "right behavior" of individuals, while others deal with public policy. On a closer look, particularly in areas such as market behavior, ethics becomes complicated because it involves the right behavior at different levels, a system that might be called *ethical federalism*. This concept, introduced earlier as Green federalism, recognizes that ethical obligations entail an interaction of the ethics of governments, private institutions, and individuals. Indeed, ethical norms at one level may differ depending upon how well the other levels are performing.

Personal ethics are at the foundation of any ethical structure—these are norms of how we as individuals should treat each other. At an intermediate level are the ethics of private institutions such as corporations or universities. At the highest level are the ethics of the state—how our governments should enact and enforce the laws and regulations in order to promote the well-managed, good society. A central complication for ethics is that all these interact since the ethics of the individual may depend upon whether people live in a well-governed state (such as, perhaps, Sweden) or in terrible tyranny (such as Nazi Germany).

This discussion primarily addresses ethics at the level of institutions and individuals. However, to proceed we need to consider the governmental ethical structure. To make this manageable, assume that we live in a well-managed society.[2] (This approach is described in chapter 3.) Recall that a well-managed society is designed to advance the welfare of its members and has four key pillars. These include laws to define property rights and contracts so that people can interact fairly and efficiently; effective markets where people can engage in transactions for private goods; laws, regulations, expenditures, and taxes to correct important externalities and provide public goods; and corrective taxation and expenditure to help ensure the fairness of the distribution of income, wealth, and power.

Ethical Actions: Negative, Positive, Neutral

We typically have multiple interactions with others on a daily basis. Some take place in the market (such as buying a pair of shoes), while others are more direct (such as driving down the street).

How can we judge the ethical status of our actions? The approach followed here is a "consequentialist" criterion based on the external effects of our actions. Under this approach, *an act is ethically positive if it improves the welfare of others, ethically negative if it harms the welfare of others, and ethically neutral if it has no effect on others.* Some acts are ethically ambiguous if they have mixed effects on others, but that complication is put aside for now.

We can first apply this definition to our day-to-day market transactions. One of the major results of modern economics is the invisible-hand principle, which refers to the efficiency of competitive markets. This is put eloquently in *The Wealth of Nations*: "It is not from the benevolence of the butcher, the brewer, or the baker that we expect our dinner, but from their regard to their own interest."

The idea behind the invisible-hand principle is that, in a well-functioning market economy, the pursuit of profits by firms and satisfaction by consumers leads to an efficient allocation of resources. When I engage in buying or selling, this will generally improve the economic welfare of those I trade with. *The invisible-hand principle means that a person's market transactions in a well-regulated society are ethically positive or neutral because they generally raise or leave unaffected the welfare of others.*

When it applies, the invisible-hand principle greatly simplifies our ethical lives because it implies that we can go about our daily economic activities without worrying that we are routinely hurting other people. All that is needed for ethical behavior is to act as responsible members of the market community: to earn and pay but not to steal or cheat.

An additional and little-appreciated aspect of the invisible hand is its informational efficiency for ethical behavior. We do not need to know anything about the butcher or the brewer or the baker to be confident that our actions are ethically neutral or positive. A well-functioning price system economizes on the need to gather mountains of information to act ethically. This point will become particularly important when we consider how to deal with our externalities.

I will close this section by emphasizing the "half-full" nature of the invisible-hand principle that is being invoked. Economists have devoted many a book to analyzing qualifications and pointing out exceptions. For our purpose, the important qualification is the presence of negative externalities. Other significant problems will arise because of uncertainty, unfairness of the distribution of income, macroeconomic distortions, and individual irrationality. I skate over these qualifications not because they are negligible but to emphasize the central ethical implications of market transactions in a well-regulated economy.

A Well-Managed Society and Individual Ethics

Starting with the idealized world of Adam Smith, we now move to the actual world of invisible-hand failures. This book is about Brown phenomena and Green policies. In reality, people are colliding with each other in negative externalities and sometimes in ways that are life- or even society-threatening. Whether the interactions involve actual or virtual collisions, societies need ways to reduce the harms from the externalities of pollution, global warming, and war.

As mentioned above, among the requirements of the governance of a well-managed society is the need to analyze and regulate important externalities. Take the example of driving a car. A well-managed society will deal with the automotive externalities through multiple laws and customs, including speed limits, stop lights, traffic fines, and liability laws that govern behavior.

Here is where the ethical federalism of a well-managed society enters. Individuals may treat driving as ethically neutral as long as government regulations have internalized the externalities. I need to obey the rules of the road and drive carefully, but I do not need to weigh the ethics of every stop sign. I am not an expert in traffic engineering, so I leave it to engineers to decide where stop signs should go. I could second-guess them, but the informational and legal burdens of second-guessing traffic engineers are sufficiently large that I generally keep within the boundaries of society's traffic laws. As with driving, so with many other well-regulated externalities.

Moving to the example of pollution, a well-managed society requires that pollution externalities be internalized. This can be done by regulations or pollution taxes or liability rules, and the best approach will be determined by technical factors. An important example is climate change caused by carbon dioxide (CO_2) emissions. Economists believe that the most effective way to slow climate change is to have a carbon price that fully reflects the social cost of CO_2 emissions.

So here is the key ethical point, using carbon emissions as an example. Assume that countries have imposed a universal carbon tax that approximates the social cost of emissions. As a result, all goods have embedded carbon charges that reflect their carbon footprints. The presence of the carbon charge can replace worries about our personal carbon footprint. When carbon is properly priced, we can go about our daily lives confident that our personal carbon emissions are in the ethical neutral zone. We would be buying carbon emissions in the same way we buy shoes and bread.

Figure 19-1 shows the trade-off between our own and others' welfare for private goods like bread or for correctly priced externalities. Each axis shows the economic welfare for self and for all others in some common metric like dollars or bundles of goods. If I cut back on my emissions by an extra unit, the loss to me is exactly matched by the gain to others. This is the fundamental result of efficient pollution regulation.

Departures from a Well-Managed Society: Unregulated Externalities

If only the economic world were so simple, and we lived in a well-managed society in which governments and markets team up to manage the economy efficiently and fairly. Alas, we must be realistic and recognize that no society will implement perfectly all the conditions for a well-managed society.

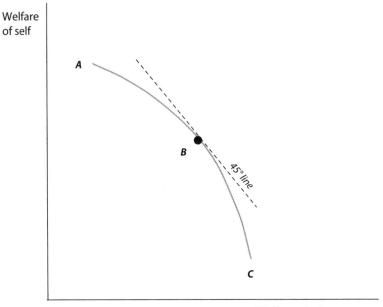

FIGURE 19-1. The curved line *ABC* represents the trade-off between own welfare and others' welfare in a well-managed society. At the market equilibrium at point *B*, the welfare of others substitutes on a one-for-minus-one basis with my own welfare.

In fact, moral philosophy studies in exquisite detail situations in which our actions may have harmful or beneficial effects on others. Economic ethics, as outlined above, holds that we should pay the full cost of our activities. Failing that, we will be causing uncompensated harms on others, which is our economic definition of unethical actions.

A possible moral principle for unregulated harmful externalities follows from this ethical view: You shall not harm others, and if you do, you should compensate them. For example, you should not damage your neighbor's car, and if you do, you should compensate your neighbor for the damage.

The damaged car is an example in which laws have internalized the ethical obligation. You are obliged to pay for the damages. However, many other cases involve costly spillovers with no obligation to pay those who are harmed. An interesting case is congestion. When

I drive to New York City on a congested highway, I am usually upset by all my wasted time. But I usually forget that I am also adding to the congestion and therefore wasting other people's time. There are no mechanisms to penalize me for wasting the time of others, and other drivers are not compensated.

What are my ethical obligations in a congested world? Should I stay home? Drive on an uncongested back road and waste hours of my time? As far as I can tell, this is one conundrum that ethicists have not touched.

By contrast, ethicists have written extensively on climate change. There is little doubt that when you drive your car, you are a contributor to climate change and that you will add a tiny bit to the damages.

What are individual ethical obligations regarding unregulated externalities? I suggest one central and surprising answer. *Our primary ethical obligation is as citizens to promote laws that correct the spillovers.* For example, we should work to ensure the enforcement of existing rules on toxic wastes or to pass laws that will slow climate change. This principle applies with special force to managers and directors of companies who are in the responsible industries. For example, automobile and energy companies should lend their weight to the political process of persuading legislatures to pass effective laws.

Beyond the rule of active citizenship, the ethics of externalities are murky because the effectiveness of actions will depend on institutional structures and technology. Let us examine some of the dilemmas and potential solutions.

The No-Regrets Policy

One interesting result comes from economics and can be helpful in thinking about how to manage our harmful spillovers or footprints (carbon footprints, congestion footprint, noise footprint, and so on). I call it the *no regrets* policy. *In the case of unregulated externalities, small reductions in our footprint have very small impacts on ourselves but large reductions in harm to others.* In other words, by taking small steps, you can reduce your spillovers, perhaps substantially, without having any regrets because there are almost no impacts on you.

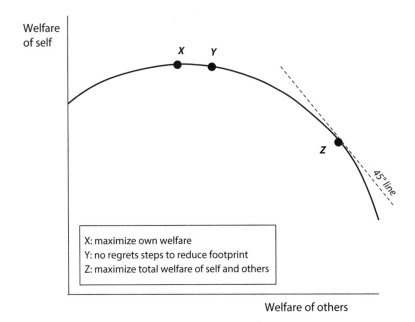

FIGURE 19-2. Impact of reducing footprint in three cases: no steps, no regrets, and complete altruism

Here is the basic reasoning using the example of air-conditioning. Suppose you like the temperature in your house to be 70°F in the summer so that is where you set your thermostat. However, on reflection, you can hardly tell the difference between 70°F and 71°F. Therefore, you turn your thermostat up to 71°F. Your welfare loss is imperceptible. However, the effect of this small change is substantial on your fuel use. A typical household would reduce its electricity use by 10%, with a substantial reduction in its carbon and pollution footprint.

Figure 19-2 shows three different stances toward the treatment of external effects on others. As in figure 19-1, each axis shows the economic welfare measure of self and of all others in some common metric. If an individual maximizes the person's own welfare and ignores the impact on others, maximization will lead the person to the top of the welfare hill at point *X*. At the very top, a tiny sacrifice will have a large effect on others and only a small impact on self, as shown at point *Y*. This important result comes because, at the top of the welfare hill, the slope is close to zero. So small changes have a tiny impact on one's own welfare but, for important externalities, they may have a large impact on the welfare of others.

If the person is a pure altruist and interested in maximizing the average welfare of everyone, that would lead to giving away much welfare to point Z. At that point, every unit of sacrificed own welfare leads to just one additional unit of others' welfare. The major point here is that in cases of important externalities, small acts of altruism can be taken with tiny own impacts but with substantial impacts on total welfare.

Here is an example to make the point. Suppose you are driving on an empty highway and encounter an elderly couple who are stranded. They explain that not only has their car broken down, but their cell phone is dead. Would you please call their friend to come and pick them up? In return for this virtually costless act of kindness, you save their day and perhaps even more. Experiments indicate that people will often make this kind of tiny sacrifice. So taking the step of moving from X to Y is a plausible act for most people.

However, suppose the conversation turns dark, and the couple say that since you have two cars, you should give them your car. They will give you a ride to your house, where you can continue your day's work, and they can share in your good fortune of having two cars. This scenario is unlikely to appeal to most people, and moving from X to Z is an unlikely act. So here is a central finding about environmental ethics relying on the no regrets approach:

You can make a substantial improvement to the general welfare and reduce your external impacts on others by taking small steps to reduce your externality footprint.

Ethics of Externalities: The Case of Climate Change

Major externalities like air pollution or climate change will not be solved without strong legal and regulatory measures. But ethics goes beyond the law. What are our ethical obligations, as individual people, or firms? Should we take the low-cost, no regrets steps to reduce our impacts described in the last section? If so, how low is low?

This question has been thoughtfully addressed by the Oxford moral philosopher and economist John Broome. His bottom line is that you should take personal steps to be "carbon neutral." I will examine the strengths and weaknesses of his argument in this section.[3]

Broome holds that the principle of justice is not to harm others, and if you do harm them, you should compensate those you have harmed. Broome's duty of justice runs reasonably parallel with Anglo-American law, as well as common sense. Broome puts several conditions on the requirement for compensation (or what he calls restitution). The seven conditions are that you harm someone, you are responsible for the act, the harms are serious, the act is not accidental, the act benefits you, there is no reciprocal benefit, and restituting actions are inexpensive.

Broome suggests that greenhouse-gas emissions meet all seven conditions. Moreover, in his view, restitution can be accomplished by completely offsetting emissions and thereby having a zero-carbon footprint. Specific offsetting actions according to Broome include energy conservation, growing trees, and purchasing offsets from a commercial company.

Here is an example. Suppose I decide to drive round trip from New Haven to Boston, which is about 270 miles. If I go to a carbon footprint calculator, I find that the emissions are about 200 pounds of CO_2. If I look at popular offset programs, such as Terrapass, I can buy offsets for about $10 per ton, so this would add about $1 to the cost of my trip. If Terrapass is effective (a point discussed below), buying their offsets seems like restitution.

However, as ethicists, we need to dig deeper. How persuasive is Broome's argument? To begin with, meeting all seven conditions would seem an Olympian hurdle. One problem with his restitution plan relates to effectiveness. Suppose you decide to conserve energy in a region that has capped emissions, such as the European Union. If emissions are capped, your energy conservation will have no impact on emissions or harms. Rather, when you limit your emissions, the emissions of others rise to offset your reductions. So total emissions are unchanged.

This example may seem overly technocratic. However, it reflects a deep issue that arises for many externalities. In most cases the effectiveness of individual actions depends upon the fine-grained technological structure of the externality. To take Broome's example of offsets, I noted how your offsets would have no impact on future damages in a region where emissions are capped. However, if emissions are not capped, as in the United States, offsets would indeed reduce emissions and reduce future damages.

Another issue that Broome skates over is the question of costs and benefits. One of his conditions is that the restitution is "inexpensive." Would this suggest the no regrets policy as a standard? Or would it mean taking any action that does not bankrupt me?

How to Reduce Your Carbon Footprint?

Many people today are concerned about global warming and would like to make personal efforts to reduce their carbon emissions. They have read about different programs and proposals. What should they do? The dilemmas with carbon offsets will illustrate the issues faced in dealing with our unregulated spillovers.[4]

The first possibility is to live in a cave, although that is not a top suggestion because not everyone would find it attractive or even feasible. A second and sounder approach is to reduce our personal carbon emissions. This involves buying energy-efficient cars and equipment, using energy-efficient light bulbs, using renewable sources where possible, and weatherizing our homes. But even the most diligent person cannot reduce the carbon footprint to zero.

The question is therefore whether to buy "offsets" that reduce carbon emissions somewhere else so that your total emissions after offsets are small or even zero. For example, a company is planting trees in the Amazon, and those trees absorb 1 ton of CO_2. By purchasing the offset, you are effectively reducing emissions by that ton.

So far, so good. But how do you know that emissions are actually reduced? You must ensure that the company is sound, that it actually is planting the trees, that someone is verifying the plot of land,

and that the trees will be there permanently. These seem difficult but manageable.

However, the most intractable part is to ensure that the emissions reductions are "additional." Perhaps the landowner was going to plant the trees anyway. Or the trees were going to be planted in a neighboring plot and were just moved to your area.

There are groups that attempt to verify all these aspects of offsets. But many economists worry that, in a world where there are no emissions limits, ensuring that the offsets are additional is virtually impossible. Devising better ways to measure the effectiveness of offsets is a high priority.[5]

The bottom line on individual actions to solve externalities is this: Individual actions, such as those recommended by the no regrets approach, are inexpensive and effective. However, they are by their nature small and therefore unlikely to provide an adequate solution to major problems. They are also inefficient because they are uncoordinated and will end up with different levels of action for different people and firms. Returning to the central point above, these endeavors cannot serve as a substitute for strong collective action by governments.

The Informational Burden of Green Ethics

Most ethical decisions involve simple issues and require little new information. The Old Testament commandments include do not steal, kill, or commit adultery. These acts are easy to understand. Today, we might ask: Did you hit a parked car? Did you rob a bank? Hit someone over the head with a lead pipe? The ethics and law here are straightforward. It is hard to argue if you are caught on video with a gun and an empty bag in the branch bank.

Impersonal activities involved in Green ethics are more complicated. What is the ethics of adding to congestion? Or adding to air pollution? Or eating meat? For these, the linkage between actions and consequences is more remote.

Climate change is an important example of where the ethics are burdened by weighty informational deficits. Before I did the calculation reported above, I did not know what my CO_2 emissions would be for the trip to Boston. I am similarly unsure about my total carbon footprint. On its website, the offset firm Carbonfund informs me that our family footprint is 24 metric tons per year. But it did not ask about my income, my air travel, or the size of my house. The estimate is worthless and designed mainly to sell offsets.

However, suppose we decide to cover our carbon footprint by purchasing carbon offsets. We might want to know whether the offset companies actually reduce emissions. Returning to the example of Terrapass, I looked at its website. One part of its portfolio is wind power. It owns part of an Oklahoma wind farm called Big Smile Wind Farm at Dempsey Ridge. The complication is that Oklahoma has a voluntary renewable portfolio standard that recommends that 15% of its power come from renewable sources like wind. Big Smile would qualify for this goal. We would worry that Big Smile power would simply displace the power generated from another wind farm. If so, there would be no aggregate reduction of emissions. But since it is voluntary today, perhaps it is additional, although that could change if mandatory standards are enacted, as is the case in many states.

Going Forward on Green Ethics

The following is a cautious summary of the discussion. It cannot capture the subtle and deep discussion by moral philosophers in this area; rather, it attempts to simplify the dilemmas to cover everyday activities of busy but responsible citizens.

First, if we follow the consequentialist framework of the Spirit of Green, market transactions in a well-managed society are ethically neutral or positive. This result enormously simplifies the ethics of everyday market transactions.

Second, the major dilemmas of environmental ethics concern activities with harmful external effects that have not been internalized by law or custom. Such effects are market failures, wherein the price I pay for my activities is significantly below their social costs.

Third, the primary responsibility of individuals and organizations is to work for collective actions to correct the externalities.

Collective actions are much more effective than uncoordinated private actions. Such collective action might be through providing reliable information from scientists and firms, through developing better antipollution laws, through social insurance, and through other social mechanisms.

Fourth, a special and useful case is the no regrets policy. When we encounter unregulated externalities, small reductions in our externality footprint have tiny impacts on ourselves but can produce large reductions in the harms to others. However, these steps are necessarily small and cannot substitute for strong collective action.

As a final cautionary note, we have found that undertaking individual actions to reduce externalities (such as reducing carbon emissions) is often complicated by institutional, technical, and informational factors that may impede effective action. Lack of knowledge makes it difficult for individuals to determine the most efficient manner to deal with their spillover actions.

20

Green Corporations and Social Responsibility

One of the major developments in the modern era is environmental, social, and corporate governance (ESG). This activity goes by different names: corporate social responsibility, socially responsible investment, and sustainable finance. But for this book, we stick with the ESG label, which is currently widely used.

ESG refers to the three central measures of the societal impact of a company. It is the projection of Green philosophy into the business world. The basic idea is that corporations are more than just money-making machines that buy steel, produce cars, and fight tooth and claw to enrich their owners. Rather, corporations are increasingly viewed as members of a society that have certain legal, economic, and ethical obligations. ESG goes beyond just obeying the law. It involves voluntary actions in which a business monitors and ensures its compliance with the spirit of the law, with ethical standards, and with national or international business norms. Moreover, ESC recognizes that profits—which are a central goal of business—are sometimes a misleading compass and need to be corrected.

This chapter deals with corporate social responsibility, which involves the activities of a corporation itself. The next chapter turns to the role of socially responsible investment, or how financial investments can be viewed from a Green vantage point.

Corporate Social Responsibility

ESG is an important development of the last half century. The basic idea is that corporations are powerful economic and political entities and must recognize their broader impacts if they are to continue to receive support in democratic societies. Three different approaches to corporate management are often emphasized: the shareholder view, the stakeholder view, and the societal view.[1]

The *shareholder view* holds that the sole responsibility of corporations is to maximize profits, or more generally to maximize shareholder value. This view is influential in some financial and economic circles and will be addressed in the next section.

The *stakeholder view* broadens out the first one by moving from shareholders to stakeholders. Stakeholders are those who are heavily affected by corporate actions. There are internal stakeholders such as stockholders, employees, and customers and external stakeholders such as communities. Hence, this view holds that firms should balance shareholder value against the impacts on other stakeholders.

The *societal view* emphasizes the place of corporations in the broader society. This differs mainly from the stakeholder view in broadening the scope of stakeholders to include the entire society. So this approach views corporations primarily as citizens.

The span of definitions is extremely wide—from the narrow economic interests of owners to the welfare of the entire society. Most advocates of corporate responsibility hold that companies should do more than the bare minimum. However, whether their duties are limited to their most affected stakeholders or to the broader public interest, or some mixture of the two, is a subject of debate.

Social Responsibility to Maximize Profits

With this overview behind us, let us begin with the extreme view of corporate responsibility, forcefully articulated by Milton Friedman.

Friedman argued that conventional corporate social responsibility is actually irresponsible. He is often viewed with suspicion as a free-market fundamentalist, but let us look at what he actually said: "There is one and only one social responsibility of business—to use its resources and engage in activities designed to increase its profits so long as it stays within the rules of the game, which is to say, engages in open and free competition without deception or fraud."[2]

Friedman's views have been formulated into a strategy called *value maximization*, which is widely taught in business schools. One of the most influential proponents of value maximization is Michael Jensen of Harvard Business School. In his formulation, value maximization states that managers should make all decisions so as to increase the total long-run market value of the firm. Total value is the sum of the values of all financial claims on the firm, including equity, debt, and other claims.[3]

Jensen's argument parallels Friedman's but adds a few wrinkles. He begins with Friedman's social role of profits: "200 years of work in economics and finance indicate that social welfare is maximized when all firms in an economy maximize total firm value."[4] Jensen also argues strenuously against broadening corporate behavior to include "stakeholder" interests. He argues that the concept is too vague to serve as an objective of management. As such, it allows management too much discretion to invest in their favorite projects and divert funds from their rightful owners.

This Chicago-school formulation has some implicit assumptions that are challenged by advocates of broader views of ESG. In reality, the conditions for the invisible-hand principle are unlikely to be met. The most important failures are market power such as that of Google or Facebook, externalities such as pollution, and inequalities of income and wealth. Other important problems would arise because of the absence of markets (particularly for the future), uncertainty, macroeconomic distortions, and irrational individual decision-making.[5]

We might think the Friedman view would be appropriate for the atomistic, perfectly competitive Farmer Joneses of the economic textbook. Jones needs to keep a sharp eye on profits or go bankrupt.

However, modern corporations are not tiny specks on the landscape. They have great discretion in managing their operations. Moreover, with the growing globalization and deregulation, governments have diminished control over the activities of companies. The ESG movement can be interpreted as a reaction to the growing autonomy of companies; it asks that companies better govern themselves. Among the broad goals that firms should consider are environmental impacts, labor practices, educational practices, transparent reporting, and adequate returns on investments.

Friedman holds that corporate responsibility requires staying "within the rules of the game." What exactly are Friedman's rules? Exactly what game is he referring to? Do the rules involve only obeying the letter of the law and staying out of jail? Or do they also involve concerns with externalities that have not been written into law? And should companies ignore pecuniary externalities, such as the serious harms to workers and communities with plant closings? In reality, some important externalities (such as those involving carbon dioxide [CO_2] emissions) are not internalized in the United States. Moreover, corporations have broad discretion to engage in political activities and to influence scientific research and public opinion in areas affecting their profits. Hence, the guideline of staying within the rules of the game is too vague to be useful.[6]

Moving beyond Friedman's theory, some argue that public corporations are *required* to maximize profits. What are the legal constraints? In the United States, it is generally held that directors of publicly owned corporations like Amazon or General Motors must act in the best interests of the corporation. But this does not mean single-minded profit maximization. This was clearly stated by the U.S. Supreme Court:

> While it is certainly true that a central objective of for-profit corporations is to make money, modern corporate law does not require for-profit corporations to pursue profit at the expense of everything else, and many do not do so. For-profit corporations, with ownership approval, support a wide variety of charitable causes. So long as its owners agree, a for-profit corporation may

take costly pollution-control and energy-conservation measures that go beyond what the law requires.[7]

However, there is one strand of value maximization worth emphasizing— the goal of avoiding *short termism,* or focusing on short-term objectives. There are always temptations to focus on short-horizon objectives such as quarterly profits or earnings per share. Often, managerial incentives are based on these short-horizon factors, which then gives management incentives to engage in myopic decisions. Jensen and others emphasize that *enlightened value maximization* encourages managers to think of the role of stakeholders creatively over the longer term, but again with the aim of maximizing the market value of the firm.

ESG and Legal Incompleteness

Governments cannot regulate every societal ailment. Perhaps the cost of regulation is greater than the damage from the ailment. Or, in many cases, private interests have more political heft than those who represent the public interest. In political systems, short-term goals, such as winning the next election, shortchange the future. Also, to be realistic, legislatures have limited time to write all the necessary laws.

Because laws cannot cover every market or societal failure, the result is *legal incompleteness.* This denotes a situation in which laws do not encompass all possible contingencies that may arise. William Landes and Richard Posner explain this syndrome:[8]

> The limits of human foresight, the ambiguities of language, and the high cost of legislative deliberation combine to assure that most legislation will be enacted in seriously incomplete form, with many areas of uncertainty left to the courts.

Faced with legal incompleteness, there are two strategies. One is to fill in the gaps to make the legal structure more complete. This would focus on areas where collective action is the most important and least controversial. For example, given the rapid rise of cybercrime and violations of privacy, which were clearly unanticipated by legislatures a century ago, improving the legal structure in these areas is a high priority. Similarly, dealing with global warming requires collective action at national and international levels of the hierarchy.

However, we must recognize that laws are likely to remain incomplete in many areas. ESG plays an important role in filling the void left by legal incompleteness.

The issues raised by legal incompleteness were analyzed by Christopher Stone in his magnificent book *Where the Law Ends: The Social Control of Corporate Behavior.*[9] His starting point, in much the same spirit as Green thought, concerns the limitations of the law when the invisible hand of markets fails to keep corporations within socially desirable boundaries. In a democratic society, if a majority of the political actors believes that the present laws are inadequate to limit the activities of corporations, they can pass tougher laws. But, as we emphasized in the chapters on Green politics, democracies are imperfect: governments are slow, reactive, and often unrepresentative. In the era of globalization, national governments have limited jurisdiction over global markets. Stone argues that, *because the law does not and cannot guide societies in a complete manner, corporations need to be restructured to fill the gap between social objectives and an incomplete legal system.*

Thus, the starting point of Stone's view of ESG is to redesign corporations so that they can remedy gaps in the legal systems. Governments should, for example, ideally tax or cap CO_2 emissions to slow global warming. If governments fail to curb CO_2, it becomes the social role of corporations to take policies that limit emissions.

Suppose we accept Stone's view that companies should fill the void where markets are inefficient, and governments have failed to act. Where does this lead? The ideas from social responsibility cover such a broad range of potential activities that it is difficult to get a foothold here. Where does ESG fit within this vast terrain? How much should companies spend? Should they stay close to home or go to the neediest areas? Who are the stakeholders, and which ones are most important?

Finally, whenever we impose ESG constraints on corporations, we need to compare the inefficiencies of corporations alongside the inefficiencies of the markets. To make this concrete, think about the potential impact of ESG on your favorite good or service. Would you have companies devote more of their resources to ESG at the cost of slowing innovation? Is ESG more important than improving smartphones? Or faster Wi-Fi services? Or introducing more effective vaccines? The task of ESG is to ensure that the economy continues to produce high-quality goods and services while also reducing the spillovers that attend those production processes.

Thus, the basic insight here is that ESG should step in where markets and governments both fail to ensure the efficient provision of important private and public goods and services.

Corporate Responsibility to Deal with Externalities

Given the vast array of possible targets for ESG, I will focus on externalities as a way of identifying appropriate ESG activities. Recall that externalities occur when the costs of an activity spill over to other people, without those other people being compensated for the damages.

As we discussed in chapter 4, externalities come in two varieties: technological and pecuniary. Most of the discussion in this book, and indeed in economics, is about technological externalities. These are spillovers, like pollution, where the interaction occurs *outside* the marketplace.

A different set of spillovers is pecuniary externalities, which are effects that take place indirectly *through* the marketplace. Pecuniary spillovers occur when economic actions affect the prices and incomes of other people.

A pecuniary externality occurs when a company closes a lumber plant in Maine and buys less expensive lumber from Canada. Similar decisions may lower the cost of building houses and improve the living standards of millions of people. However, the plant closure destroys the jobs of hundreds of workers whose incomes fall sharply. This interaction occurs *through* markets, and is called pecuniary, rather than *outside* markets like pollution or congestion.

ESG might be targeted toward both kinds of externalities. The reason is that corporations may need to step in because the political process does not adequately protect and compensate those who are harmed. The lack of social protections might come because of scientific uncertainty or political obstruction or international free riding or a weak social safety net.

ESG becomes particularly important in the presence of technological uncertainty. How harmful are DDT, asbestos, sulfur dioxide, CO_2, low-level radiation, and ozone-depleting chemicals? Companies producing goods that contain or emit these substances are often legally responsible for their impacts. They are also, or ought to be, the most informed about the effects.

So here is the summary: When a company pollutes its local community in a legal fashion or harms its workers through labor practices or plant shutdowns, these are areas where ESG would most naturally be an issue. Our revised and preferred definition of ESG for this book is therefore the following:

> Environmental, social, and corporate governance, or ESG, involves alleviating pecuniary or technological externalities caused by the firm. The most relevant are impacts on stakeholders such as employees and local communities, ones that have particularly grave societal impacts, and ones in which the company has specialized and privileged knowledge.

Confronting the Trade-Off between ESG and Profits

The central issue in ESG is the potential conflict between profits and socially responsible behavior. Some strategies are *win-win* (or W, W) behavior that benefits society (the first win) and increases profits (the second win). A firm with a long horizon might understand that some of the ESG activities are in fact profitable in the long run; perhaps they enhance the firm's reputation and increase sales or lower costs. No responsible director would argue against (W, W) activities that decrease short-run profits but increase long-run profits. Such corporate actions are *enlightened profit maximization*,

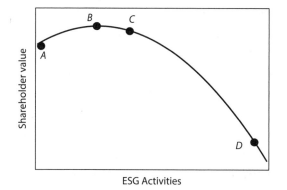

FIGURE 20-1. Four strategies for corporate responsibility

in which ESG is a sophisticated and profitable business practice. There are no real trade-offs here.

Most dilemmas in ESG involve "win-lose" (W, L) behavior. These are actions that increase the economic or social welfare of nonowners but reduce profits and shareholder value. A firm that reduces emissions more than required, or keeps a domestic factory going a little longer, or improves working conditions above the competitive standard may hurt its long-run bottom line—so this is a (W, L) situation.

Figure 20-1 illustrates the point. This shows the shareholder value or profits for different levels of ESG or Green activities. Point A has zero ESG. If it acts in this manner, the company is both socially and financially irresponsible. Point A is actually less profitable than point B, which deploys several profit-augmenting ESG activities, such as enhancing employee health or labeling Green products.

Moving to B consists of (W, W) activities that are in the interests of the firm's stockholder value. These activities attract customers, hold off boycotts, lower the cost of capital, and provide good public relations. It is hard to argue against actions that are (W, W) between A and B. Indeed, if Friedman were persuaded that curbing local pollution or training the workforce was profitable, he would surely favor such ESG activities. However, he would point out that they are profitable and leave off any ESG motivation.

Consider ESG activities that go a little beyond B to C and are of the (W, L) category. Perhaps lowering pollution in the host city below the letter of the regulations imposes some costs, but it promotes the public health of its workers and their families. As a responsible citizen, the corporation goes beyond enlightened self-interest to point C, with a public benefit that offsets the loss in profits. The annual rate of return for shareholders is slightly lower.

However, point C also illustrates the no-regret policy. Perhaps the firm reduces emissions a little more than required, or keeps a marginal plant open, or operates a day care center for employees, or sets up a pension plan with incentives for worker saving. Each of these might cost a little in profits, but they can make a substantial contribution to the welfare of stakeholders.

We can also envision going off the cliff with ESG, as with point D. Perhaps the company decides to buy a failing business or go to Mars. These actions drive the firm into unprofitable territory. Few managers or stakeholders would defend this strategy since the company would soon be out of business.

Principles for Unprofitable ESG

Should companies engage in unprofitable (W, L) activities? The literature on ESG is tangled on this subject. In contemplating whether ESG might lower profits, a proponent will list six reasons why it will not actually be unprofitable and therefore will fall into (W, W) behavior. However, some ESG activities are genuinely (W, L) and lower corporate profits. What principles can be laid down on win-lose behavior?

The literature has few clear guidelines, but I would offer the following three suggestions. A first guideline for Environmental, Social, and Governance is that the activities should pass a *social* benefit-cost test even if they fail a *private* profit-cost test. Therefore, if the firm has a pollution externality that causes $100 of social damage, it might undertake to spend up to $100 to reduce that pollution. But it would make no sense to spend $200 of the firm's (and society's) resources to reduce $100 in social damages. This first guideline clearly removes many activities from the list of socially responsible activities.

However, a vast number of projects can undoubtedly pass a social-cost benefit test. These would include subsidizing education in Africa or building medical clinics in poor neighborhoods. How would the firm choose among the array of possible ESG projects? Two further guidelines can help companies.

The second guideline is that firms should concentrate their resources on areas where they have comparative informational or economic advantages. For example, firms often have specialized knowledge about the dangers involved in their products or processes. They can usefully study their activities to determine the harms and take steps to prevent them. An example of a company that went this route is DuPont, which pushed for the introduction of substitutes for ozone-killing chlorofluorocarbons (CFCs). DuPont probably suffered some damage to its bottom line, but its actions made the phaseout of CFCs much more successful. Too often, firms have the opposite inclination. Automobile companies dragged their feet on new technologies, such as airbags in cars, that later proved highly successful. One of the most damaging cases of withholding information has been Facebook, which has systematically profited on its customer information, lied about its activities, and helped bad actors affect elections and poison public attitudes.

A third guideline is to focus on those ESG activities that primarily benefit stakeholders but select projects within that group with a high social benefit-cost ratio. Some examples might be early childcare and health programs for the firm's workers. Firms might consider revising their implicit contracts with workers to improve workers' economic and social conditions. A firm might be particularly reluctant to close plants that are marginally profitable. The stakeholder approach is one that views the corporation as a small society rather than as a profit-making machine. Corporations should be active participants in this minisociety and particularly attentive to workers, communities, and long-term customers.

These guidelines rest on the recognition that firms know their business and neighborhoods but have little expertise in determining what is in the public interest. Their specialization is their market. An automobile company knows how to design airbags and how to reduce emissions efficiently, but its managers generally have no training in public health, cost-benefit analysis, and the comparative value of alternative health and safety regulations. So this final point emphasizes that ESG should involve areas where the company has special expertise or responsibilities.

The Ninth Circle of Corporate Irresponsibility

Most of the focus in business schools and academic writing is on corporate responsibility: do this, do that, measure this, report that. I close this chapter with an alternative approach that looks at corporate *irresponsibility*. One widely used rating is the KLD Social Ratings Database. This rating has categories of irresponsible behavior, including environmental issues (such as hazardous waste releases), corporate governance (such as excessive executive compensation), and production in controversial sectors (alcohol, tobacco, and fossil fuels). As is the case in many measures of ESG performance, there is no easy way to produce an overall index from such data.

Many of the most egregious forms of corporate misconduct involve providing misleading or fraudulent information about the company's own products and processes. Such behavior is a risk to the public that the company is uniquely positioned to understand. This kind of fraud is worse than simple thievery. It is particularly harmful because those who are most informed are using their knowledge to mislead the public.

When Dante wrote *Inferno*, he described the ninth circle of hell as the deepest one, where treachery resides and where hosts betray their guests.

> By effect of his malicious thoughts,
> Trusting in him I was made prisoner,
> And after put to death.
> (*Inferno*, Canto XXXIII)

So it is with companies in the ninth circle. We are indeed prisoners when we trust malicious companies who invite us as guests into their showrooms. These firms know about their dangerous products, withhold that knowledge, subvert science to advance their narrow commercial interests, and put their guests to death.

A recent egregious example is Volkswagen, which not only hid the emissions of its diesel automobiles and fabricated the results but designed equipment to falsify the results. It did this to save money on the production of purportedly "clean" diesel engines. How many people died as a result? How many people bought VW cars believing they were Green? While these questions are unanswered, the behavior was not only illegal but belongs in the ninth circle of corporate irresponsibility.

Here is a sample of companies residing in the ninth circle:

- Volkswagen, for designing equipment to falsify emissions testing
- Philip Morris, for hiding research showing the lethal nature of smoking
- ExxonMobil, for suppressing the science of climate change and funding climate deniers
- Johns Manville, for knowledge and denial of the dangers of asbestos for years before lawsuits ferreted out the truth
- Purdue Pharma, for misrepresenting the addictive qualities of OxyContin
- Facebook, for misrepresenting the treatment of personal information and selling it to vendors around the world, including Russians who sowed discord in other countries

Students who look to affect their universities' investment policies would do a service by systematically finding and evaluating these worst of the irresponsible activities.

Final Words

A tour through the literature on corporate responsibility finds a confusing medley of themes and vantage points. However, if we stand back and look at the landscape, four key findings emerge.

First, there is much discussion but little agreement on ESG. There are no standard ESG metrics nor are there accepted way to aggregate the different measures into an overall metric. Companies often receive (or claim to receive) high ESG marks, but their public reports are often superficial, so individuals cannot make their own judgments about performance. Moreover, many of the companies that rate ESG do not make their rating systems public, so we cannot judge what the ESG scores actually represent. Actual measures of ESG live in a dense fog.

A second point to emphasize, which pervades virtually every area discussed in this book, is that companies should avoid short termism. In other words, companies should be structured to take a broad and long view of what improves long-run profitability and shareholder value. This involves structuring managerial incentives to avoid focus on short-run returns. Companies are well advised to take a broad view of their corporate culture as well as the communities in which they live. Devoting resources to improving the lives of their workers and the reliability of their products can be wise long-term investments.

A third point is to remember the no-regrets principle. When companies help correct externalities, they can make substantial contributions to stakeholders and society with only small impacts on profits. This principle applies in many areas where the entity is optimizing its behavior, and therefore small deviations from the optimum can have substantial external impacts with small internal impacts.

Additionally, it would be useful for those who review company performance to get a good measure of the resources that are devoted to ESG and to separate out spending that is public relations. We should remain skeptical of corporate social spending that is devoted to building goodwill. If you walk by Lincoln Center in New York, you will see the David H. Koch Theater. Support for the arts may deflect criticism from the Koch brothers undermining environmental regulations, but it does little to clean the environment or meet environmental standards.

Finally, companies have an especially important role in today's technologically complex economy to provide accurate information about the potential risks of their products and processes. This is where companies have deep knowledge. They have the responsibility to be honest with their customers and not to hide dangers or mislead government regulators. The worst companies knowingly kill people through dangerous or faulty products, and these companies deserve the most severe sanctions.

21

Green Finance

We have explored the difficulties of ensuring that corporations behave in socially responsible ways through codes of conduct and external monitoring. An interesting alternative is for the *owners* of public corporations to insist that these corporations engage in socially responsible behavior. This is another application of the principles of environmental, social, and corporate governance (ESG) discussed in the last chapter. It sometimes goes by the name of socially responsible investments, or ethical investment, but increasingly is combined with other areas into ESG.

What is ESG for finance? A simple statement is that these are financial investments that include environmental, social, and corporate governance factors in decisions. ESG can be a powerful tool in inducing firms to behave in Green directions because owners have the legal power to determine corporate decisions. They who pay the piper call the tune.

ESG has grown rapidly in recent years. According to one survey, $12 trillion (or about one-quarter) of professionally managed U.S. assets applied ESG criteria in their investments in 2018.[1] The major areas of concern were climate change, tobacco, conflict risk, human rights, and transparency.

Closer to my home, universities have been urged to invest their endowments in socially responsible corporations. In an earlier era, some universities urged firms not to locate to South Africa, and other universities divested of tobacco stocks. Today, a vocal movement advocates that universities sell firms who produce or distribute fossil fuels because they contribute to global warming.

The movement to promote Green investments faces many of the issues raised for socially responsible corporations. What is social responsibility? How can we define and measure it? Does it involve primarily ensuring long-run profitability and avoiding short termism? Is it designed to include the externalities from firm decisions, such as climate change? Do Green investments penalize investors? And perhaps most important, can it be effective?

What Are Socially Responsible Investments?

ESG for finance closely parallels the definitions of socially responsible corporations analyzed in the last chapter. However, there is one major difference: Green finance looks particularly at what a firm *produces*, while corporate responsibility mainly looks at *methods* of production.

Here is an example. Today, ExxonMobil produces and sells fossil fuels. Analysts would ask whether Exxon, as a responsible corporation, is engaged in fair labor practices, discloses its products and environmental impacts, and has ambitious goals for its carbon footprint. You might be impressed to learn that ExxonMobil won several awards in the last few years as the best company in social responsibility in several areas.

However, in the case of finance, ExxonMobil is in the crosshairs of many advocates of ethical investment because it produces and sells oil and gas and thereby contributes to climate change. Another group of excluded firms are those producing guns, tobacco, alcohol, and military weapons. These might be called "sinful firms," not because they act in sinful ways (they might, like ExxonMobil, be model firms and follow the law) but because they sell products that have harmful effects.

Why Be Socially Responsible? Enlightened Profits

Investors face the same dilemmas and trade-offs as socially responsible corporations. One goal is to choose companies that act in the long-term interests of their owners; this goal requires that firms maximize shareholder value, considering societal trends. A representative statement about an ESG mission is the one adopted by the giant pension and money manager TIAA.

> As providers of capital, long-term investors have among the most to lose if markets deteriorate and asset prices fall. Therefore, it is critical that such investors use their influence and leverage to promote good corporate governance and effectively functioning markets. Our participants and clients expect us to be stewards of their savings and to help provide for their financial security.[2]

There is no altruism in this statement—no concern for the externalities of the investments—just long-term financial returns.

Another giant in the investment industry is the California public pension system (CalPERS), which manages more than $300 billion in assets. It recently told its money managers that they would be required to incorporate ESG goals into their strategic planning. Here is a statement to investment managers from a 2015 presentation:

> CalPERS must consider risk factors, for example climate change and natural resource availability, which emerge slowly over long time periods, but could have a material impact on company or portfolio returns.[3]

A careful reading of this directive indicates that climate change should enter the analysis because it affects portfolio returns, not because company actions cause climate change. Therefore, this analysis clearly uses a justification of long-run profitability in its ESG strategy.

Why Be Socially Responsible? Public Purposes

While the two giant pension funds just examined focus on financial returns, other ethical investors include the public impacts of their investments.

To explore ESG with broader social goals, let us begin with individual investors. Unlike corporations, individuals face no legal or economic constraints on their altruistic goals and will face no shareholder protests. If a radical utilitarian like Peter Singer wants to give away virtually all his money to equalize marginal satisfaction around the globe—that is his money. If the libertarian philosopher Robert Nozick counters that he has no duty to help others, no one can force him to put his pension into a Green fund. So, subject to the constraints of law, individuals can devise their own investment philosophy.

Corporations operate under tighter constraints. In an earlier era, it was not clear that corporations were allowed to make charitable contributions. It is now settled that as a matter of law corporations may make unlimited charitable contributions. They must, however, reckon with their charters, directors, and owners. Hence, while corporations may take actions that reduce shareholder value, perhaps devoting 1% of their profits to community activities, the owners would undoubtedly revolt if a firm gave away 99% of its profits.

The ESG policies of most investment firms are usually fuzzy and provide little information or guidance to investors. We saw typical language in the statements of TIAA and CalPERS above. You will almost never find a clear statement about the extent to which a company will accept a penalty in its returns in order to promote social justice.

Yale's Ethical Investment Policy

While most financial managers state their ESG objectives vaguely, my home university speaks clearly about its approach. I will rely on the investment philosophy of Yale University for this discussion because I am familiar with the local landscape, and Yale has clearly articulated its goals.

The core of Yale's policies was developed in a pathbreaking report written by a masterful Yale Law School professor, John Simon, and two colleagues. The guidelines have two key premises. The first is that "maximum economic return will be the exclusive criterion for selection and retention of the university's endowment securities."

Second, there are narrow cases in which the university will *sacrifice return* on its portfolio for social objectives.[4] The university will take steps from shareholder resolutions to divestment if the company is causing a "social injury." These are activities of a company that have an "injurious impact . . . on consumers, employees, or other persons." Importantly, it particularly includes "activities which violate, or frustrate the enforcement of, rules of domestic or international law intended to protect individuals against deprivation of health, safety, or basic freedoms."

Therefore, there are two criteria for exclusion from the investment portfolio. First, the investment would cause social injury (in other words, it would be some kind of externality), and second, the activity would violate or impede domestic or international laws protecting individuals. The second restriction is unusual in ESG guidelines in having a narrow focus on illegal, rather than unethical, activities.

Over the years, Yale has taken actions regarding apartheid in South Africa and against tobacco companies, as well as oil and gas companies operating in South Sudan. It has taken cautious steps on climate change, encouraging the disclosure of emissions and the analysis of the risk impact of climate change on investment performance. Given the limited scope of these actions, it is unlikely that the returns on the Yale endowment (which has the highest long-run return of any major university) have suffered.

The unusual aspect of the Yale rule is the stance on legal pollution and other widespread externalities. These might be sanctioned if they were the fault of only a single company. However, for industry-wide externalities, the report is concerned that a single company would be at a competitive disadvantage if it alone were targeted for sanctions or divestment. In those areas, government action would be required, such as banning assault rifles or limiting emissions.

In cases like industry-wide pollution, where government action is necessary, the university will limit its activities to "communicate with the management of the company to urge it to seek necessary action from the appropriate government agencies." This rule would exclude divestment of individual fossil-fuel companies but require the companies to work to seek strong governmental policies.

It is on this point—concerning industry-wide social injuries—that Yale's guidelines part ways from those of many of its students, and indeed from other universities, who would like to see divestment of companies *legally* producing fossil fuels, guns, and tobacco.

Socially Responsible Investment in Practice

We can look at some of the most important ESG funds to understand their philosophy. Vanguard's Social Index Fund is one of the largest. Here is its investment philosophy:

> The Index excludes companies that have violations or controversies related to (1) environmental impact, (2) human rights, (3) health and safety, or (4) labor standards, or that (5) fail to meet criteria related to diversity. Also excluded from the Index are companies that are involved with (1) weapons, (2) tobacco, (3) gambling, (4) alcohol, (5) adult entertainment, or (6) nuclear power.

Note that Vanguard includes legal as well as illegal activities. It is not clear why some sectors are targeted and not others. The fund does hold companies producing oil, automobiles, chemicals, banking services, and other sectors that might be objectionable. Also, some might wonder why Vanguard penalizes nuclear power producers since they reduce overall emissions of greenhouse gases.

A second important example is TIAA-CREF's Social Choice Equity fund. Here is its statement:

> The Fund's investments are subject to certain environmental, social, and governance ("ESG") criteria. . . . All companies must meet or exceed minimum ESG performance standards to be eligible for inclusion in the Fund. The evaluation process favors companies with leadership in ESG performance relative to their peers.

Firms engaged in "production and sale of alcohol, tobacco, military weapons, firearms, nuclear power, and gambling products" are

TABLE 21-1. Comparison of shares of the largest companies in a market fund with shares in the TIAA-CREF Social Choice Fund and the Vanguard Social Choice fund

Company	Total Market	TIAA	Vanguard
Microsoft	3.86%	4.10%	5.77%
Apple	3.56%	4.00%	5.98%
Amazon	2.63%	2.40%	
Facebook	1.62%		2.44%
Berkshire Hathaway	1.42%		
JPMorgan Chase	1.34%		2.11%
Alphabet	2.64%	3.00%	4.04%
Johnson & Johnson	1.23%		1.87%
Procter & Gamble Co.	1.10%	1.50%	1.64%
Visa	1.09%		1.65%
Exxon Mobil Corp.	1.01%		
AT&T	0.99%		
Bank of America	0.92%		1.40%
Home Depot	0.90%	1.30%	1.37%
Intel Corp.	0.88%	1.30%	1.33%
Verizon	0.88%	1.30%	
Mastercard	0.88%		1.32%
Walt Disney	0.82%		1.21%
UnitedHealth	0.81%	1.00%	1.27%
Merck	0.78%	1.20%	1.19%

Source: The holdings in each of the portfolios were obtained from the websites of the funds as of November 2019.

penalized but not automatically excluded. The exact definition of the minimum ESG standards is not explained.

Green Portfolios in Practice

We can go beyond rhetoric to action by looking at the actual portfolios of different Green funds. Two of the largest ESG portfolios are those of Vanguard and TIAA, whose criteria were just cited. Let us compare the holdings of these two Green portfolios with a standard index fund of the entire U.S. market.[5] Table 21-1 shows the shares of the top 20 companies in the market portfolio (those with the largest market capitalization). It also shows the shares of those 20 companies held by two important ESG funds.

Some important points emerge. First, individual stocks generally are more heavily weighted in Green portfolios because they omit a significant fraction of the total market. For example, Microsoft is upweighted from 3.9% to 5.8% in Vanguard Social.

Second, the choices might appear puzzling. Take Amazon and Facebook. Both companies have average ESG scores. However, TIAA includes Amazon but excludes Facebook, while Vanguard does the opposite. ExxonMobil has a low ESG score because it produces fossil fuels, although it had a high score until 2019. The exclusion of banks and credit-card companies from TIAA but not Vanguard is a puzzle. And the exclusion of the House of Mouse (Walt Disney) from TIAA might surprise TIAA shareholders.[6]

One interesting finding from this simple exercise is that the exclusions appear arbitrary and depend upon the tastes of the investment fund managers. From a financial point of view, excluding a substantial fraction of the top firms can lower diversification and returns. The arbitrary nature of the exclusions reflects the lack of any systematic way of measuring company ESG performance that we have seen repeatedly in this and the prior chapter.

What Is the Cost of a Green Portfolio?

While some ethical propositions are absolute, most issues require weighing the costs against the benefits. Most investors will ask how much it will cost to exclude certain companies or sectors.

Standard investment advice today is to hold a broadly diversified portfolio of securities. For example, a typical index fund might hold the largest 500 companies identified as the Standard and Poor 500 (S&P 500). It is a "passive" fund because it requires no one to decide on the companies. This lowers the costs of operating the fund.

A broad portfolio is also diversified and reduces the exposure to the bad fortunes of a single company. For example, if you held shares of Facebook, it would have had a volatility of more than double that of the S&P 500.

Hence, there are two costs of holding a Green portfolio. First, you would need to pay someone to decide on the exclusions, and

TABLE 21-2. Returns on Green funds

Portfolio	Return	Expenses	Net return
Market	6.00%	0.04%	5.96%
Vanguard Social	5.93%	0.18%	5.75%
TIAA-CREF Social	5.81%	0.22%	5.59%
Average ESG	5.80%	0.93%	4.87%

Note: This shows the impact on the risk or expected return of excluding stocks in the two Green funds and a hypothetical high-cost fund. The last column indicates that the return penalty can be substantial.

second, it would exclude companies or sectors and therefore be less diversified. I will use a simple example to illustrate the costs of Green investment.[7]

Table 21-2 shows the expected impact on annual returns of limiting the portfolio for the TIAA and Vanguard Green funds as well as for the average of ESG funds.[8] The first column shows the expected return (corrected for risk). Social choice funds have slightly lower expected returns because they have less diversification. For most funds, the return penalty is likely to be small, between 0.1% and 0.2% per year.

However, the major cost is the higher expense ratio, shown in the second column. The total loss is around 0.3% per year for Vanguard and TIAA but much larger for the average ESG fund in the last row. If you are not careful, you might well lose 1 percentage point of your 6-percentage-point return.

These results are meant to illustrate the impacts of ESG exclusions on returns. Studies looking at the actual returns of ESG funds find varying results. One reason is that many studies look at ex post or historical returns, which include purely random and one-time factors. Green funds typically lose more than the small amounts shown in Table 21-2, but the reasons are unclear. The poor performance may have arisen because of a poor choice of investments by managers of Green funds.

Additionally, some studies look at the impact of excluding individual stocks, and for a single stock, the impact will indeed be small. For example, an advocate of "Fossil Free U" might do the same analysis and calculate the impact of excluding ExxonMobil on the return of a portfolio of 500 largest companies. The same calculation used in table 21-2 would show that the expected return would decline from 6.000% to 5.997%, and this is clearly trivial. The reason the return penalty is so low is that ExxonMobil is only a tiny fraction of the portfolio. If the fund excludes energy companies, banks, automobile companies, utilities, companies with a presence in questionable countries like China, and chemical companies, by contrast, the return penalty would be much higher.

Investment Strategy for Green Investors

What are the lessons for Green investment? Here are the points that emerge.

First, whether or not you are interested in Green investments, always look to companies that take the long view. Shun companies in which management is self-serving and shortsighted.

Second, we can apply the no-regrets principle to Green finance. If the portfolio is optimized to begin with, then small exclusions from the portfolio will have negligible impacts on returns. So, if a fund excludes only a few companies or a small sector, it will have only a small penalty in long-run returns.

Third, if you decide to put a Green tint on your portfolio, choose your targets sparingly. If you want to have the cleanest of investments, you are likely to take a significant punishment on your returns. This means that you should look at what the fund actually holds. If the exclusions are extensive, or hard to understand, or do not conform to your philosophy, then perhaps look elsewhere.

And pay close attention to expenses. If you are not careful, these can eat up your earnings. One of the worst funds was FundX Sustainable Impact, with an annual expense ratio of 2.1%. Some funds will even add a sales charge. You would do much better with Vanguard's 0.20% expense ratio with a 0.00% sales charge.

PART V

Global Green

22

Green Planet?

Most of the facets of Green discussed up to now operated at the personal, local, or national level. However, some of the most intractable and risky externalities are global. We discussed one important global ailment, pandemics, in an earlier chapter. This and the next chapter survey global Green as represented by global warming.

Climate Change as a Global Externality

Climate change is a particularly thorny externality because it is global. Many critical issues facing humanity today—global warming and ozone depletion, COVID-19, financial crises, cyberwarfare, and nuclear proliferation—are similarly global in effect and resist the control of both markets and national governments. Such global externalities, whose impacts are indivisibly spread around the entire world, are not new, but they are becoming more important because of rapid technological change and the process of globalization.[1]

Global warming is the Goliath of all externalities because it involves so many activities. It affects the entire planet for decades and even centuries, yet none of us acting individually, or even as nations, can do much to slow the changes.

Global externalities have long challenged national governments. In earlier centuries, countries faced religious conflicts, marauding armies, and the spread of pandemics of smallpox and the plague. In the modern world, the older global challenges have not disappeared, as we see with the COVID-19 pandemic, while new ones have arisen—including not only global warming but others such as the threat of nuclear proliferation, drug trafficking, and international financial crises.

Further reflection will reveal that nations have had limited success with agreements to deal with global economic externalities. Two successful cases include handling international trade disputes (today primarily through the World Trade Organization) and the protocols to limit the use of ozone-killing chlorofluorocarbons. The study of economic aspects of environmental treaties has been pioneered by Columbia University economist Scott Barrett. He and other scholars believe these two treaties were successful because the benefits far outweighed the costs and because effective institutions were created to foster cooperation among nations.[2]

Governance is a central issue in dealing with global externalities because effective management requires the concerted action of major countries. However, under current international law, there is no legal mechanism by which disinterested majorities of countries can require other nations to share in the responsibility for managing global externalities. Moreover, extralegal methods such as armed force are hardly recommended when the point is to persuade countries to behave cooperatively.

It must be emphasized that global environmental concerns raise completely different governance issues from national environmental concerns, such as air and water pollution. For national public goods, the problems largely involve making the national political institutions responsive to the diffuse national public interest rather than concentrated national private interests. For global public goods, the problems arise because individual nations enjoy only a small fraction of the benefits of their actions. In other words, even the most democratic nations acting noncooperatively in their own interests would take minimal actions because most of the benefits spill out to other nations. It is only by designing, implementing, and enforcing *cooperative multinational policies* that nations can ensure effective policies.

This chapter discusses the scientific and economic background to climate change. The next chapter explores global mechanisms (what I call climate clubs or compacts) to deal with the lack of incentives to manage global externalities.

The Changing Science of Climate Change

If you read the newspaper, listen to the radio, or read Twitter, you are virtually certain to encounter stories about global warming. Here is a sample from a variety of sources:

"The last decade was the warmest on record."
"The concept of global warming was created by and for
    the Chinese in order to make U.S. manufacturing
    non-competitive."
"Polar bears could disappear within a century."
"The Greenland ice sheet has experienced record melting."

Clearly, global warming is getting a lot of attention today. And just as clearly, people disagree about whether it is real, whether it is important, and what it means for human societies. What should the interested citizen conclude from these conflicting stories? And if the answer is that global warming is real, how much does it matter? Where should our concerns about global warming rank among the other issues we face, such as persistent inequality, pandemics, and nuclear proliferation?

The short answer is that global warming is a major threat to humans and the natural world. It is the ultimate challenge for Green policies, threatening to turn Planet Earth into Planet Brown.

I have used the metaphor that climate change is like a vast casino. By this, I mean that economic growth is producing unintended but perilous changes in the climate and Earth systems. These changes will lead to unforeseeable and probably dangerous consequences. We are rolling the climatic dice, and the outcome will produce surprises, some of which are likely to be perilous. The message in these chapters is that we can put down the climatic dice and walk out of the casino.

Global warming is one of the defining issues of our time. It ranks along with pandemics and economic depressions as a force that will shape the human and natural landscapes for the indefinite future. Global warming is also a complex subject. It spans disciplines from basic climate science to ecology and economics, and even includes politics and international relations.

Climate Basics

A few chapters in this book cannot hope to cover the vast scope of climate change. Rather, this discussion will highlight the major issues involved, explain why climate change threatens the planet, and show how these relate to the overall Green philosophy in this book.[3]

The beginning of our understanding lies in earth sciences. Climate science is a dynamic field, but the essential elements have been developed by earth scientists over the last century and are well established. The ultimate source of global warming is the burning of fossil (or carbon-based) fuels such as coal, oil, and natural gas, which leads to emissions of carbon dioxide (CO_2). Gases such as CO_2 are called greenhouse gases (GHGs). They accumulate in the atmosphere and stay there for a long time.

Higher atmospheric concentrations of GHGs lead to surface warming of the land and oceans. The initial warming is amplified through feedback effects in the atmosphere, oceans, and ice sheets. The result includes changes in temperatures as well as in temperature extremes, precipitation patterns, storm location and frequency, snowpacks, river runoff, water availability, and ice sheets. Each of these will have profound impacts on biological and human activities that are sensitive to the climate.

Past climates—varying from ice-free conditions to Snowball Earth—were driven by natural sources. Current climate change is increasingly caused by human activities. The major driver of global warming is the emission of CO_2 from the burning of fossil fuels.

CO_2 concentrations in the atmosphere were 280 parts per million (ppm) in 1750 and have reached over 410 ppm today. Models project that, unless forceful steps are taken to reduce fossil-fuel use, concentrations of CO_2 will reach 700–900 ppm by 2100. According to climate models, this will lead to warming averaged over the globe in the range of 3–5°C by 2100, with significant further warming after that. So, unless there are strong efforts to curb CO_2 emissions sharply, we can expect continued accumulations of CO_2 emissions in the atmosphere—and the resulting global warming with all its consequences.

Is this all a fantasy of scientists who are looking for funding for their pet projects? Such a cynical and misguided view not only insults the talented people who have labored in this field but also overlooks the powerful evidence they have provided. Figure 22-1 shows one critical piece of evidence here, the record of atmospheric CO_2 concentrations over the last 800,000 years. You can see the seesaw of concentrations associated with the ice ages. Cold periods were those in which CO_2 declined sharply (probably because it went into the deep oceans), while warm periods led to large CO_2 releases. Concentrations varied from lows around 170 ppm to highs around 280 ppm in the preindustrial period. During the most recent ice age, global temperatures were about 5°C lower than today, and atmospheric CO_2 concentrations were at their lowest point, 180 ppm.

Then, around 1750, as humans began clearing forests and burning fossil fuels, CO_2 concentrations headed up. Concentrations passed the 800,000-year record around 1950 and by 2020 were 410 ppm. Carbon-cycle models indicate that the elevated levels result from industrial emissions, with about half of all emissions from the last century remaining in the atmosphere—and likely to stay there for a century or more.

The accumulating CO_2, along with other GHGs, is leading to rising temperatures and other accompanying climatic effects. Global temperatures have risen more than 1°C over the last century. If emissions continue unabated, climate models suggest that global temperatures will rise another 2–4°C by the end of the century. Some areas, such as the Arctic, will see much sharper temperature increases.

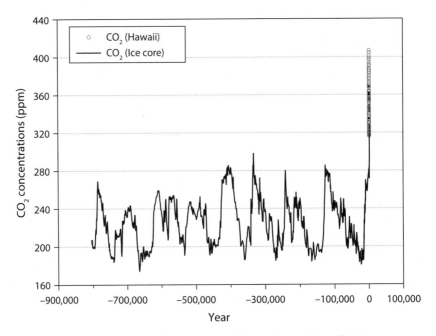

FIGURE 22-1. CO_2 concentrations from ice cores and historical record through 2020
The longer solid line comes from ice cores in large ice sheets such as Antarctica. The dots starting in 1957 are instrumental records from Hawaii.

But temperature is only a small part of the impacts, many of which are imperfectly understood. Among other impacts are drying in the midcontinental regions, more intense storms, smaller glaciers and snowpacks, perhaps more widespread wildfires, and changing monsoonal patterns.

Figure 22-2 shows a reconstruction of global temperatures using Antarctica ice-core data for the last half-million years. The temperature at present is normalized at 0°C. The line with dots shooting up at the far right shows a projection of future temperature increases if there are no policies to slow climate change. If global warming continues unchecked, future temperatures will soon surpass the historical maximum of the last half-million years.

Rising temperatures are not the major concern about the impacts of climate change. More important are the effects on human and natural systems with regard to storms, giant ice sheets, and monsoonal systems. A central concept in analyzing impacts is whether a

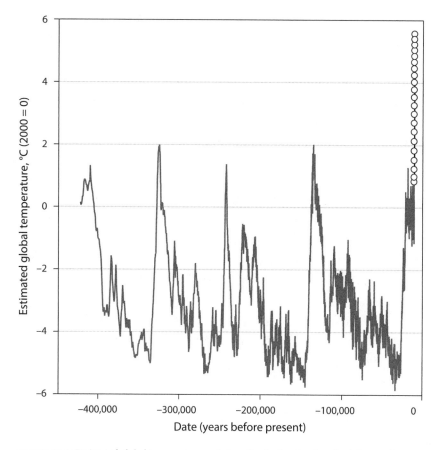

FIGURE 22-2. Estimated global temperature variations for the last four hundred thousand years (*solid line*) along with model projections for the next two centuries (*circles*)

system can be managed. The nonagricultural sectors of high-income countries are highly managed, and this feature will allow these sectors to adapt to climate change at a relatively low cost for at least a few decades.

However, many human and natural systems are unmanaged or unmanageable and are highly vulnerable to future climate change. While some sectors or countries may benefit from climate change, most countries are likely to be significantly disrupted in sectors closely tied to climate-sensitive physical systems. The potential damages will probably be most heavily concentrated in low-income and tropical regions, such as tropical Africa, Latin America, coastal communities, and the Indian subcontinent. Vulnerable systems include rain-fed agriculture, seasonal snowpacks, coastal communities impacted by sea-level rise, river runoffs, and natural ecosystems. There is potential for serious impacts in these areas.

Scientists are particularly concerned about *tipping points* in the earth's systems. These involve processes in which sudden or irreversible changes occur as systems cross thresholds. Many of these systems operate at such a large scale that they are effectively unmanageable by humans with existing technologies. Four important global tipping points are the rapid melting of large ice sheets (such as Greenland and Antarctica); large-scale changes in ocean circulation, such as the Gulf Stream; melting of the permafrost; and major changes in monsoonal patterns. These tipping points are particularly dangerous because they are not easily reversed once they are triggered.

The best evidence indicates that the impacts of climate change will be nonlinear and cumulative. For example, the first 1°C or 2°C of warming is unlikely to have massive disruptive effects on agriculture, particularly if warming is gradual and farmers can adapt their technologies. However, as global warming passes the 3°C or 4°C mark, the combination of changes in temperature, precipitation, and water availability is likely to highly disrupt most agricultural systems.

The Climate Deniers

The science and economics of major environmental issues is vigorously debated and sometimes denied by those who cause the problems and whose interests would be adversely affected by mitigating policies. We saw that when Rachel Carson warned the world about the dangers of DDT and other pesticides, she was targeted as enemy number one by Big Chemicals. Similarly, energy companies, particularly those producing or selling fossil fuels, see their profits threatened if strong climate policies are established. The most damaging participants are politicians who argue against Green policies because of ideology or campaign contributions. Companies have the money, but politicians have the votes and the power.

I have studied climate science for decades and find it solid and convincing. But there are skeptics. Many people misunderstand the issues. A few influential politicians sow doubts about the validity of mainstream climate science. Affected industries undermine the science and exaggerate the costs of policies to slow warming. Here are some examples of the contentious dialogue:

> FROM PRESIDENT DONALD TRUMP: "The concept of global warming was created by and for the Chinese in order to make U.S. manufacturing non-competitive."
>
> THE TITLE OF A BOOK BY U.S. SENATOR JAMES INHOFE: *The Greatest Hoax: How the Global Warming Conspiracy Threatens Your Future*
>
> DR. WILLIAM HAPPER (SEE BELOW): "I believe that more CO_2 is good for the world, that the world has been in a CO_2 famine for many tens of millions of years."
>
> FROM A KEY ADVISER TO RUSSIAN PRESIDENT VLADIMIR PUTIN: "No link has been established between carbon dioxide emissions and climate change."

The list could go on and on. While these debates seem amusing distractions, they pose serious challenges because of their impact on public opinion. It is worth looking into these claims to test their validity.

The media desires "fairness," so often an established theory will be "balanced" by some far-out idea. This has been the case for climate change. We find today a small and vocal group of contrarian scientists who argue that the consensus on climate change is poorly grounded and that policies to slow warming are not warranted.

To explain how such contrarian views are propagated, I will take the case of a 2012 article by "sixteen scientists" in the *Wall Street Journal* titled "No Need to Panic about Global Warming." Dissenting scientists here are not typically active researchers in the field but are influential because they carry the mantle of science and often have made important contributions in other areas. It is useful to look at this statement because it contains many of the standard criticisms.

The basic message of the article asserts that the globe is not warming and that CO_2 is not harmful. I will analyze two of their claims as typical of the contrarian viewpoint.

1. The first claim for contrarians is that the planet is not warming. The 16 scientists wrote, "Perhaps the most inconvenient fact is the lack of global warming for well over 10 years now."

It is easy to get lost in the tiniest details here. Just because the stock market went down today does not mean that it does not generally rise. It will be useful to look at the record of actual temperature measurements. Our best measures show that global mean temperature has risen 1.3°C since 1900, with an accelerating trend since 1980.

Moreover, climate scientists have moved way beyond global mean temperature in looking for evidence of human-caused climate change. Scientists have found several indicators that point to a warming world with humans as the major cause. These include melting of glaciers and ice sheets; changes in ocean heat content, rainfall patterns, atmospheric moisture, and river runoff; stratospheric cooling; and the shrinking of Arctic sea ice. Those who look only at global temperature trends are like investigators using only eyewitness reports and ignoring fingerprints and DNA-based evidence. Yet the contrarians continue to repeat their claims using outmoded techniques and data.

2. One of the strangest claims of contrarians is the second argument: "The fact is that CO_2 is not a pollutant." What might this mean? Presumably, it means that CO_2 is not by itself toxic to humans or other organisms within the range of concentrations that we are likely to encounter, and indeed higher CO_2 concentrations may be beneficial.

However, this is not the meaning of pollution under U.S. law or in standard economics. The U.S. Clean Air Act defined an air pollutant as "any air pollution agent or combination of such agents, including any physical, chemical, biological, radioactive . . . substance or matter which is emitted into or otherwise enters the ambient air." In a 2007 decision, the Supreme Court ruled on the question:

> Carbon dioxide, methane, nitrous oxide, and hydrofluorocarbons are without a doubt "physical [and] chemical . . . substance[s] which [are] emitted into . . . the ambient air." . . . Greenhouse gases fit well within the Clean Air Act's capacious definition of "air pollutant."[4]

In economics, a pollutant is a form of negative externality—that is, a by-product of economic activity that causes damages to innocent bystanders. The question here is whether emissions of CO_2 and other GHGs will cause damages, large or small, now and in the future. Virtually all studies of the impacts of rising concentrations of CO_2 and the accompanying earth-system changes have concluded that there are net damages, that the damages are large, and that the damages rise sharply for warming greater than 1°C. In short, CO_2 is indeed a pollutant in the sense that it is a damaging side effect of economic activity.

Other claims of contrarians range from the absurd (it is a hoax created by the Chinese in order to make U.S. manufacturing non-competitive) to the abstruse (clouds will save the globe from catastrophic warming).

Economics of Climate Change

We move now from science to economics. Economists have focused on strategies to slow climate change. The most promising is *mitigation*, or reducing emissions of CO_2 and other GHGs. Unfortunately, this approach is expensive. Studies indicate that it will cost in the range of 2 to 6% of world income (roughly, $2 trillion to $6 trillion annually at today's level of income) to attain international climate targets, even if mitigation is undertaken in an efficient manner. While some miraculous technological breakthroughs might conceivably be discovered that can reduce the costs dramatically, experts do not see them arriving in the near future. New technologies—particularly for energy systems that have massive investments in capital such as power plants, structures, roads, airports, and factories—take many decades to develop and deploy.[5]

The economics of climate change is straightforward. When we burn fossil fuels, we inadvertently emit CO_2 into the atmosphere, and this leads to the harmful impacts just discussed. As explained elsewhere in this book, such a process is an externality, which occurs because those who generate the emissions do not pay, and those who are harmed are not compensated. One major lesson from economics is that unregulated markets cannot efficiently deal with extensive harmful externalities. Unregulated markets will produce too much CO_2 because there is a zero price on the external damages of CO_2 emissions.

Economics points to one central and all-important truth about climate-change policy. This truth is so central that it must be stated and restated. For any policy to be effective, it must raise the market price of CO_2 and other GHG emissions. Putting a price on emissions corrects for the underpricing of the externality in the marketplace. Prices can be raised by putting a regulatory tradable limit on the amount of allowable emissions (cap-and-trade) or by levying a tax on carbon emissions (carbon tax).

A central lesson of economic history is the power of incentives. Take the example of land values. Where land is scarce and land prices are high, such as on the island of Manhattan, people build smaller dwellings and go high into the sky. Where land prices are low, such as in southern New Mexico, people worry little about the cost of the land and spread out their houses and barns.

Applying that to our subject, we can ask how to use incentives to slow climate change. Here, the incentive must be for everyone to replace their current fossil-fuel-driven consumption with low-carbon activities. Making this change requires the actions of millions of firms and billions of people spending trillions of dollars.

The most effective incentive to induce the transition is a high price for carbon. Raising the price of carbon will achieve four goals. First, it will signal to *consumers* which goods and services are carbon-intensive and should therefore be used sparingly. Second, it will provide data to *producers* about which inputs are carbon-intensive (such as coal and oil) and which are low carbon (such as natural gas or wind power), thereby inducing firms to move to low-carbon technologies. Third, it will give market incentives for *inventors, innovators, and investment bankers* to invent, fund, develop, and commercialize new low-carbon products and processes. Finally, a carbon price will economize on the *information* required to undertake all these tasks.

Economists have extensively studied the major questions of climate-change policy: How sharply should countries reduce CO_2 and other GHG emissions? What should be the time profile of emissions reductions? How should the reductions be distributed across industries and countries? What policy tools are most effective—taxes, market-based emissions caps, regulations, or subsidies? Here are some of the findings.

It is tempting to set climate objectives as hard targets based on climate history or ecological principles. A common target is to limit global temperature increase to 2°C; more recently, scientists point to a limit of 1.5°C as the upper bound if we are to protect many biological processes and avoid dangerous tipping points. However, these aspirational goals may be infeasible given the current trajectory of emissions, as well as the slow pace of actions in taking strong policies.

Economists often advocate an approach known as cost-benefit analysis, in which targets are chosen by balancing costs and benefits. Because the mechanisms involved in climate change and its impacts are so complex, economists and scientists have developed computerized *integrated assessment models* to project trends, assess policies, and calculate costs and benefits. Here are some of the major findings:[6]

- Policies to slow emissions should be introduced *as soon as possible*.
- A second and surprising finding is the importance of harmonizing climate policies. This requires equalizing the marginal costs of reducing emissions everywhere. Equivalently, in a market context that means the carbon price should be equal in every sector and every country.
- Effective policies should have the highest possible *participation*; that is, the maximum number of countries and sectors should be on board as soon as possible. Free riding should be discouraged.

- Finally, an effective policy is one that *ramps up gradually*—to give people time to adapt to a high-carbon-price world, to give firms a signal about the economic environment for future investments, and to tighten the screws increasingly on carbon emissions.

Most experts agree on these central principles—universal participation, equalizing marginal costs or carbon prices in all uses in a given year, full participation, and increasing stringency over time. However, experts disagree on the stringency of policies. I have worked on models that suggest a current carbon price in the range of $40 per ton of CO_2, rising over time. This policy would lead to eventual warming of around 3°C above preindustrial levels.

However, the most ambitious policies of limiting temperature change to 2°C would require much higher carbon prices, near $200 per ton of CO_2 in the near term. Yet other prices would be consistent with other temperature trajectories, participation rates, and discounting. A lower price is appropriate if costs are low, participation rates are high, and the discount rate on future economic impacts is high. A higher price would apply for high costs, low participation rates, and low discounting.

However, whether the goal is policies that keep temperatures near 2°C or 3°C or 4°C, we must be realistic and realize that the world is not close to attaining those goals. Effective policies have not been introduced, either in any major country or for the world as a whole. Compared to a target for current carbon prices of $40 per ton of CO_2, the actual global carbon price is close to $2 per ton in 2020. Carbon prices in the United States and most other countries are virtually zero, so there is a huge gap between reality and global aspirations.

Why have *global* policies on climate change been so ineffective compared to *many national environmental* policies (for pollution, public health, and water quality as examples)? Why have landmark agreements such as the Kyoto Protocol and the Paris Accord failed to make a dent on emissions trends? The difficulties that arise for global public goods are discussed next, along with potential solutions.

Climate Compacts to
Protect the Planet

Climate change is the ultimate Green challenge because it is a *global externality*. As we discussed earlier, global externalities are different from other economic activities because the economic and political mechanisms for dealing with them efficiently and effectively are weak or absent. The result, as we will see, is that only the tiniest of steps have been taken to slow climate change. The present chapter reintroduces and develops a radical proposal of mine—a climate compact or club—that can potentially overcome the formidable obstacles raised by nationalism and free riding.

The Syndrome of Free Riding

One major reason for the slow progress in reducing global warming is the tendency for countries to seek their own national welfare. The Trump administration highlighted a policy of "America First," but other countries have similar tendencies. Moreover, when actions do not spill over the border, countries are well governed when they put their citizens' benefits first rather than adopting policies of narrow interests that lobby for protectionist tariffs or regulatory relief.

However, nationalist policies that seek to maximize the interests of a country at the expense of other countries—sometimes called *beggar thy neighbor policies*—are a poor way to resolve global problems. Noncooperative nationalist policies for tariffs, ocean fisheries, war, and climate change lead to outcomes where all are worse off.

Some contests are zero-sum games, as when nations compete in the Olympics. Others are negative-sum games, as when nations go to war. However, many global issues are cooperative games, in which the sum of nations' incomes or welfare is improved if countries refrain from nationalistic policies and take cooperative policies. The most important examples of cooperation are treaties and alliances that have led to a sharp decline in the lethality of battle deaths (look back to figure 14-1). Another important example discussed in the chapters on Green politics is the emergence of low-tariff regimes in most countries (see figure 14-4). By removing barriers to trade, all nations have seen an improvement in their living standards.

Alongside the successful outcomes lie a string of global failures. Nations have failed to stop nuclear proliferation, overfishing in the oceans, littering of space, and pandemics. In many of these failures, we see the syndrome of free riding.

Collective security is a critical national concern subject to free riding. Some nations—particularly ones surrounded by friendly and peaceful neighbors—inevitably contribute very little to international efforts to secure peaceful resolution of disputes. For example, the North Atlantic Treaty Organization (NATO) has for seven decades successfully protected its members against attack. Each country contributes by spending its domestic resources on the common agenda of military preparedness. But within this successful structure, many small countries free ride on the activities of its largest member, the United States. Hence, the United States in 2016 spent $664 billion, or 72% of the total. Many countries spend only a tiny fraction of their gross domestic product (GDP) on defense, Luxembourg being the extreme case with only $0.2 billion, or less than .5% of its GDP. Countries that contribute little to a multiparty agreement get a free ride on the costly investments of other countries.

Free riding is a major hurdle in the solution of global externalities, and it is at the heart of the failure to deal with climate change. No single country has an incentive to cut its carbon dioxide (CO_2) emissions sharply. Moreover, if there is an agreement, nations have a strong incentive not to participate. If they do participate, there is further incentive to miss ambitious objectives. In game theory, the outcome is a *noncooperative free-riding equilibrium*—a situation that closely resembles the current international policy environment in which few countries undertake strong climate-change policies. Nations speak loudly but carry no stick at all.

In the case of climate change, additional factors impede a strong international agreement. There is a tendency for the current generation to ride free by pushing the costs of dealing with climate change onto future generations. Generational free riding occurs because most of the benefits of costly emissions reductions today would accrue many decades in the future.

So global climate-change policies are hampered by two dimensions of free riding: the first is that countries want to rely on the efforts of other countries; the second is that the present generation is tempted to defer action for future generations to pay the bills.

This double free riding is further aggravated by interest groups that muddy the water by providing misleading analyses of climate science and economic costs. Contrarians highlight anomalies and unresolved scientific questions while ignoring the strong evidence supporting the underlying science. The obstacles to effective policies have been particularly high in the United States, where the ideological opposition has hardened even as the scientific concerns have become increasingly grave. Just to get the flavor, here is a summary of arguments against climate-change policies (in my paraphrase):

Contrarians deny that the globe is warming. When that argument fails, they claim that warming is due to natural sources. Moreover, even if the globe is warming, that is purportedly good for humans because there are so many cold regions and CO_2 is a fertilizer for agriculture. But, the argument goes, even if there might be harms, reducing emissions would wreck the economy.

Yet another issue is that policies would raise production costs and hurt exports. And so on.

A Short History of International Climate Agreements

Up to here, the discussion has focused on the science and economics of climate change along with the syndrome of free riding that tends to undermine strong international agreements. We move now to a history of the actual international negotiations on climate change.

The risks of climate change were recognized in the United Nations Framework Convention on Climate Change, ratified in 1994. That treaty stated, "The ultimate objective . . . is to achieve . . . stabilization of greenhouse-gas concentrations in the atmosphere at a level that would prevent dangerous anthropogenic interference with the climate system."

The first step to implement the Framework Convention was taken in the Kyoto Protocol in 1997. High-income countries agreed to limit their emissions to 5% below 1990 levels for the 2008–2012 budget period (with different targets for different countries). Under the Protocol, important institutional features were established, such as reporting requirements. The Protocol also introduced a method for calculating the relative importance of different greenhouse gases. Its most important innovation was an international cap-and-trade system of emissions trading as a means of coordinating policies among countries. (Recall the discussion of cap-and-trade for sulfur emissions in chapter 14.)

The Kyoto Protocol was an ambitious attempt to construct an international architecture that would effectively harmonize the policies of different countries. But countries did not find it economically advantageous. The United States withdrew very early. The Protocol did not attract any new participants from middle-income and developing countries. As a result, there was significant attrition in the coverage of emissions under the Kyoto Protocol. Also, emissions grew more rapidly in noncovered countries, particularly developing countries like China. The countries included in the Protocol accounted for two-thirds of global CO_2 emissions in 1990, but that declined to barely one-fifth of world emissions by 2012. It died a quiet death, mourned by few, on December 31, 2012. Kyoto's rules on emissions were so poorly designed that it turned out to be a club no country cared to join.

The Kyoto Protocol was followed by the Paris Accord of 2015. This agreement led to a target to limit climate change to 2°C above pre-industrial levels. The Paris Agreement requires all countries to make their best efforts through "nationally determined contributions."

For example, China announced that it would reduce its 2030 carbon intensity by 60–65% compared to 2005 levels. This would amount to an annual decrease in carbon intensity of 1.7–2.0% per year. The United States under the Obama administration committed to reducing its greenhouse gas emissions by 26–28% below the 2005 level in 2025. All these steps were undermined when the Trump administration announced that the United States would withdraw from the agreement, although that would not occur until November 2020.

An important point is that the national policies under the Paris Accord are *uncoordinated* and *voluntary*. They are uncoordinated in the sense that they do not add up to policies that, if undertaken, would limit climate change to 2°C. Moreover, while countries agree to make best efforts, there are no penalties if they withdraw or fail to meet their obligations.

Hence, the world continues to recognize the danger of climate change without adopting the necessary policies to slow or stop it. This was just the state of affairs with the first international agreements in the 1990s. This is the state of affairs today, except the world is hotter, and there are 400 billion more tons of CO_2 in the atmosphere than when the first treaty was signed.

The Effectiveness of Climate Policies

After a quarter-century of international agreements, we should step back to ask how effective past international agreements have proven to be. We can look to analyses of participation, coverage, targets, and timetables. But the real answer lies in the results, particularly

TABLE 23-1. Decarbonization of the global economy

Period	World	China	World less China
1980–90	–1.9%	–3.9%	–2.1%
1990–2000	–2.2%	–5.6%	–2.1%
2000–10	–0.8%	–0.6%	–1.6%
2010–17	–2.0%	–4.7%	–1.7%
1980–2017	–1.7%	–3.6%	–1.9%

the *carbon intensity* of production (which was the Chinese target mentioned above). This measures the trend in the ratio of CO_2 emissions to output. For example, in 2010, the United States emitted 5.7 billion tons of CO_2, and its real GDP was $14.8 trillion, which implied a carbon intensity of 0.386 tons of CO_2 per $1,000 of GDP. By 2015, carbon intensity declined to 0.328, for an average rate of decarbonization of 3.1% per year.

Carbon intensity can change through three primary mechanisms: through a change in the mix of fuels (substituting wind for coal), a change in the mix of output (low-carbon consumption like telecommunications instead of high-carbon driving), and a change in the efficiency of energy use (such as more fuel-efficient autos). Climate policies can affect each of these mechanisms.

If policies were effective, then the trend in carbon intensity should have declined sharply after, say, the Framework Convention or the Kyoto Protocol. Table 23-1 shows the rate of decarbonization for the last four decades. It is useful to look at the world, at China, and at the world less China because China has become such a large contributor. Over this period the rate of decarbonization has averaged 1.7% annually for the world, much higher at 3.6% for China, and 1.9% for the world less China.

Focus on the last column of table 23-1. As seen, there have been essentially no improvements in the global rate of decarbonization. Indeed, the trend of decarbonization is slightly slower over the last two decades than in earlier decades. Figure 23-1 shows the trend and annual data for the world less China. The three landmark years (1994

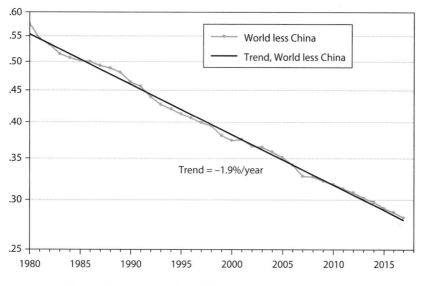

FIGURE 23-1. The trend in decarbonization, 1980–2017

for the Framework Convention, 1997 for Kyoto, and 2015 for Paris) show no breaks in the trend. While we cannot say why the trend is so persistent, it definitely shows no major change.

One reason why the emissions trend has been so persistent is that the commitments are so modest. Let us look at the commitments of the United States and China relative to the trend. For all the celebration, China's commitment is actually less than its recent trend. China would reach its target around 2030 at the current trend of decarbonization of 4% per year. Thus, China needs only to continue its current path. For the United States, the goal is slightly more ambitious. The rate of decarbonization in the United States has been 2.8% per year for the last decade, while the target would imply a rate of 3.4% per year.

A more important question is how the current rate of decarbonization would compare with trajectories that would attain the aspirational temperature targets. Figure 23-2 shows the historical and four future paths for carbon intensity using the DICE model of climate economics.[1] The baseline is a continuation of current

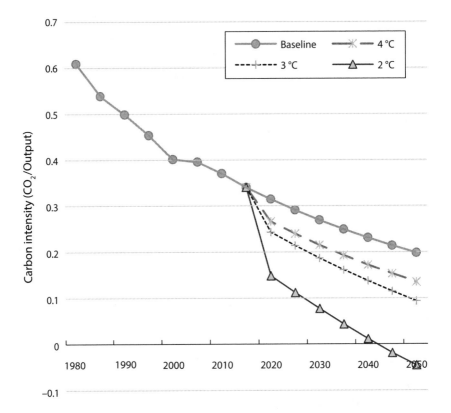

FIGURE 23-2. Carbon intensity, alternative policies

trends, which would lead to a 4+°C warming by 2100 and rise further after that.

The other three paths show the rates of decarbonization that would be associated with three different temperature limits. Note that there is an immediate sharp drop in intensity in the 2020 period as policies are introduced (starting from a world of virtually no climate policies). As an example, the current goal of 2°C would require decarbonization at about 10% per year over the next two decades (instead of the current 2%). Even more daunting is that a 2°C limit would require zero CO_2 emissions by midcentury.

The lesson here is that the policies taken to date fall far short of what is necessary to slow climate change sufficiently to meet international goals.

Climate Compacts to Overcome Free Riding

Whatever the international regime to slow climate change—whether it be a revived Kyoto approach or an updated Paris agreement—it must confront the tendency of countries to free ride on the efforts of others. Countries have strong incentives to proclaim lofty and ambitious goals . . . and then to ignore these goals and go about their business as usual. When national economic interests collide with international agreements, there is a temptation to shirk, dissemble, and withdraw.

Free riding occurs when a party receives the benefits of a public good without contributing to the costs. In the case of international climate-change policy, countries have an incentive to rely on the emissions reductions of others without taking costly domestic reductions. The failure of the Kyoto Protocol, and the difficulties of forging effective follow-up regimes, is largely due to free riding.

Canada is an interesting case. Canada was an early enthusiast for the Kyoto Protocol. It signed up for a 6% reduction in emissions and ratified the treaty. However, the Canadian energy market changed dramatically in the following years, with the rapid growth in production from the Alberta oil sands. By 2009, Canadian emissions were 17% above 1990 levels, far above its Kyoto target. Finally, in December 2011, Canada withdrew from the Protocol. There were no adverse consequences except for some scolding from environmentalists. The Canadian experience reveals a deep flaw in the Kyoto Protocol and follow-up agreements—they were toothless treaties, containing no mechanisms for enforcement. In a deep sense, participation was voluntary. It is likely that the Paris Accord will see a similar outcome.

In light of the failure of the Kyoto Protocol, it is easy to conclude that international cooperation is doomed to failure. This is the wrong conclusion. In spite of the obstacles of potential free riding, nations have in fact overcome many transnational conflicts and spillovers through international agreements. Countries enter into agreements because joint action can consider the spillover effects among the participants. These agreements are a kind of a "compact of nations" that will be described below.[2]

One particularly interesting example is the development of a free and open trading system, which we described at length in the chapter on Green politics. An important part of the success is that the World Trade Organization (WTO) has a club structure in which countries have both rights and obligations, and one of the important obligations is low tariffs. In these and other cases, the tendency toward free riding has been overcome through the mechanism of treaties.

So what is a club or a compact? Although most of us belong to clubs, we seldom consider their structure. A club is a voluntary group deriving mutual benefits from sharing the costs of producing a shared good or service. The gains from a successful club are sufficiently large that members will pay dues and adhere to club rules to gain the benefits of membership.

The major conditions for a successful club or compact include the following: a public-good-type resource that can be shared (whether the benefits from a military alliance or the enjoyment of low-cost goods from around the world); a cooperative arrangement, including dues, that is beneficial for each of the members; a rule that nonmembers can be excluded or penalized at relatively low cost to members; and a membership that is stable in the sense that no one wants to leave.

So what is the idea of a climate compact? The point is that *nations can make progress in international climate agreements if they adopt the club or compact model rather than the current voluntary model.* A climate compact is an agreement by participating countries to undertake harmonized emissions reductions, but nations would be penalized if they did not meet their obligations. The compact proposed here centers on an *international target carbon price* that is the focal provision of the agreement. For example, countries might agree that each country will implement policies that produce a minimum domestic carbon price of $40 per ton of CO_2.

One important feature of the climate compact is that it organizes policies around a target carbon price rather than emissions reductions (as with the Paris Accord and the Kyoto Protocol). One reason for focusing on prices rather than quantities is the structure of the costs and benefits. But the more important and unusual reason involves the dimensionality of the two approaches.

This point has been explored in depth by the late Harvard economist Martin Weitzman. He has shown that it would be both less distortionary and easier to negotiate a single carbon price than a set of quantity limits. The intuition is straightforward, even though the proof is difficult. In voting on a price, countries can simply negotiate for one that is near their top choice. So the United States might vote for a price close to $40 a ton, assuming that all other countries participated. For every price, each country would have a "yes, no" choice. Perhaps the price that got 50% or 75% of the votes would win.[3]

With quantities, the voting is much more complicated. There is not only a global total but also a national cap. Thus, the United States would be inclined to vote for a low global total and a high national level of emissions. Each country would do the same. There would be endless wrangling with shifting coalitions trying to benefit themselves at the expense of other groups. This difference between a single variable (the harmonized price) and many variables (the number of country caps) is a central reason why quantity restrictions are so difficult.

A key part of the compact mechanism—and the major difference from all current proposals—is that nonparticipants are penalized. While many different penalties might be considered, the simplest and most effective would be uniform percentage tariffs on the imports of nonparticipants into the compact region. The climate compact creates a strategic situation in which countries acting in their self-interest will choose to enter the compact and undertake ambitious emissions reductions because of the structure of the incentives. To understand the nature of the incentives and strategies, I discuss the application of game theory to international environmental treaties.

Both theory and history suggest that some form of sanction on nonparticipants is required to induce countries to participate in agreements with local costs but diffuse benefits. A sanction is a governmental withdrawal, or threat of withdrawal, of customary trade or financial relationships. A key aspect of the climate-compact sanctions analyzed here is that they benefit those who impose sanctions and harm those who are sanctioned. This pattern contrasts with many cases in which sanctions impose costs on the sanctioners as well as the sanctioned and thereby raise issues of incentive compatibility.

There is a small literature analyzing the effectiveness of climate compacts and comparing them to agreements without sanctions. The results suggest that a well-designed compact using trade sanctions would provide well-aligned incentives for countries to join a compact that requires strong abatement.

The international community is a long way from adopting a climate compact or any arrangement that will slow the ominous march of climate change (as seen in figures 23-1 and 23-2 above). Obstacles include ignorance, the distortions of democracy by antienvironmental interests and political contributions, free riding even among those looking to the interests of their country, and shortsightedness among those who discount the interests of the future.

Climate change and its dire consequences are the biggest threat to a Green world and pose the most daunting challenge. Global warming is a trillion-dollar problem requiring a trillion-dollar solution, and the battle for hearts, minds, and votes will be fierce.

Four Steps for Today

If climate change is the ultimate Green challenge, what can concerned citizens of the world do right now? I would emphasize four specific items to focus on.

First, people around the world need to understand and accept the gravity of the impacts of global warming on the human and natural world. Scientists must continue intensive research on every aspect from science and ecology to economics and international relations. Those who understand the issue must speak up and debunk contrarians who spread false and tendentious reasoning. People should be alert to the trumped-up claims of contrarians who find some negative results or list reasons to wait for decades to take the appropriate steps.

Second, nations must establish policies that raise the price of CO_2 and other greenhouse-gas emissions. While such steps meet resistance, they are essential elements for curbing emissions, promoting innovation and the adoption of low-carbon technologies, and inoculating our globe against the threat of unchecked warming.

Moreover, we need to ensure that actions are global and not just national or local. While politics may be local, and the opposition to strong steps to slow warming arising from nationalistic attitudes, slowing climate change requires coordinated global action. The best hope for effective coordination is a climate compact, which is a coalition of nations that commit to strong steps to reduce emissions along with mechanisms to penalize countries that do not participate. While this is a radical new proposal, no other blueprint on the public agenda holds such promise of strong international action.

Finally, it is clear that rapid technological change in the energy sector is central to the transition to a low-carbon economy. Current low-carbon technologies cannot substitute for fossil fuels without a substantial economic penalty on carbon emissions. Developing radically new and economical low-carbon technologies requires substantial public support for science and technology along with the incentive of a high carbon price. New technologies will speed the transition to a low-carbon economy and will lower the cost of achieving our climate goals. Therefore, governments and the private sector must intensively pursue low-carbon, zero-carbon, and even negative-carbon technologies.

Improved public acceptance, proper pricing, coordinated action, and new technologies—these are the steps for global Green, as well as for other important areas.

PART VI

Critiques and Final Reflections

Skeptics of Green

This book has surveyed the landscape of Green thinking in many areas. Before we turn to the summarizing chapter, it will be useful to discuss those with dissenting views. Some may think that the Green proposals in this book are too timid. Others hold that Green thinking is misguided or will wreck our economies.

Figure 24-1 illustrates the range of opinions. At the far left is the "deep Green" movement. This approach puts a heavy weight on biocentric and environmental values and a small weight on human preferences. At the far right is "muck brown," which is populated by merchants of doubt who put their profits above social welfare.

Moving toward the middle right, we find free-market environmentalism, represented especially by the Chicago school's conservative economist Milton Friedman. This approach combines a skepticism about the value of public goods and the ability of governments to regulate the economy efficiently.

Finally, we come to the Spirit of Green, which is thoroughly represented in the present book. As discussed here, the Green movement argues for the need to tilt societal laws, regulations, and values in a Green direction—to put human needs and wants at the center but to include other values as well. We touch briefly on muck brown

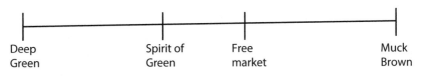

Deep
Green

Spirit of
Green

Free
market

Muck
Brown

FIGURE 24-1. The spectrum of Green

and deep Green but concentrate primarily on the contributions of free-market environmentalism.

Muck Brown

At the far right of the spectrum is muck brown. This group is populated by what can charitably be called the incentivized skeptics. These are people or companies who have economic or political motives to be skeptical of the science, economics, or ethics of Green thinking.

For example, if a company is making a hefty profit fouling the atmosphere, or even breaking the law in doing so, it will have strong incentives to argue against constraining regulations. It may round up far-out ideas or hire hungry scholars to back its activities. These groups may think it less costly to buy political support than to spend on abatement.

One prominent example is the activities associated with the Koch brothers and their companies. Koch Industries is a privately owned company holding companies with major environmental impacts and with revenues of around $115 billion in 2017. As reported by the Center for Public Integrity, one of Koch's holdings, Georgia-Pacific, is significantly impacted by the U.S. Environmental Protection Agency (EPA) decision on dioxin. They write that Koch has intervened "in various regulatory proceedings to dilute or halt tighter federal regulation of several toxic byproducts that could affect its bottom line, including dioxin, asbestos and formaldehyde, all of which have been linked to cancer."[1] According to Greenpeace, "Koch Family Foundations have spent $145,555,197 directly financing 90 groups that have attacked climate change science and policy solutions, from 1997–2018."[2]

Sometimes, political parties become identified with muck-brown opposition to environmental policies. This occurred particularly with the Republican Party in the United States after 1980. In part, the opposition stems from the funding of wealthy donors who benefit from lax environmental regulation. Additionally, in the case of the Republicans, environmental policies require active government, while the party has increasingly fought for small government and limited federal powers. Most disturbing is a cynical attitude in which the opposition to strong environmental policies turns into attacks on the basic science underlying these policies. We see absurd arguments of politicians labeling climate science a "hoax" or even "Chinese" in origin and responsibility, or labeling COVID-19 as "Kung Flu" even though it is not an influenza.

The earlier discussion on corporate responsibility argued that activities in this category sometimes descend to the ninth circle of corporate irresponsibility. These are companies or cooperating individuals who deceive the public in areas of their own specialized expertise. A recent headline case was Volkswagen, who designed elaborate schemes of deception by installing software that gave erroneous readings of emissions for their diesel cars. Debunking the arguments of incentivized skeptics is an important activity, but that is for another day. Rather, the purpose of this chapter is to engage with the serious Green skeptics.

Deep Green

At the far left is deep Green. This pole includes environmentalists and scientists who believe that the highest priority should be placed on the preservation of nature, while human values are greatly overvalued in the economy and politics. I would emphasize that there is absolutely no moral equivalence of deep Green with the muck brown approach. However, both approaches have the feature that they elevate a single value—whether private profits or the importance of nature—rather than acknowledge the need to balance competing objectives.

Deep Green encompasses a range of groups and philosophies. Among the most interesting are deep ecology, anarcho-primitivism, and ecological resistance groups.

Deep ecology was discussed extensively in earlier chapters. The idea is that all forms of life have inalienable rights, and humans have no superior claim to existence or to the use of Earth's resources. This approach generally holds that human populations and industrial activity are excessive and need to be reduced so that nonhuman life can flourish. The major precepts of deep ecology are to enhance nonhuman populations, strengthen the preservation of wilderness and biodiversity, and tread lightly (or not at all) on the planet. In its extreme versions, deep ecology is among the most radical proposals for reshaping the planet. Moreover, it is consistent with some parts of the central Green movement. But, at present and for the foreseeable future, it does not have the votes (of humans or animals) to sway elections in its favor.

Another splinter movement is *anarcho-primitivism*. This movement has roots in agrarian romanticism, such as seen in the writings of Henry David Thoreau, who celebrated, "Life consists with wildness. The most alive is the wildest." In the modern setting, anarcho-primitivism becomes alienated from modern civilization. Here is a statement by neo-Luddite Kirkpatrick Sale that captures many of the elements:[3]

> Anthropocentrism, and its expression in both humanism and monotheism, is the ruling principle of Western civilization, as to which must be opposed the principle of biocentrism. . . . Globalism, and its economic and military expression, is the guiding strategy of that civilization, to which must be opposed the strategy of localism. . . . Industrial capitalism, as an economy built upon the exploitation and degradation of the earth, is the productive and distributive enterprise of that civilization, to which must be opposed the practices of an ecological and sustainable economy.

There would be little left of human civilizations after critics have dismantled all these exploitative and degrading systems.

Additionally, deep Green includes *activists* who deploy protests, civil disobedience, and even violence to further their causes. These groups include Greenpeace, People for the Ethical Treatment of

Animals (PETA), and Earth First! Of these, Greenpeace is likely the best known. It issues reports on climate change, toxic wastes, genetically modified organisms (GMOs), nuclear weapons, nuclear power, preserving species and ecosystems, and whaling. It sometimes makes headlines when it clashes with polluters. For example, Greenpeace tried to board a Russian drill rig to protest oil drilling in the Arctic. The Russians seized a Greenpeace ship and arrested its crew as pirates. This incident raised a fiery storm of controversy and gave much favorable publicity to Greenpeace, although it did little to change the pattern of drilling in the Arctic.

Friedman and the Libertarian Tradition

The most influential critic of the Spirit of Green has been Milton Friedman (1912–2006), who was a persuasive advocate of what can be called *free-market environmentalism.* The basic idea is that free markets are not only essential to raising living standards but are also inherently Green.

The central premise of Friedman's work is the relationship between liberty and a free market:[4]

> Historical evidence speaks with a single voice on the relation between political freedom and a free market. I know of no example in time or place of a society that has been marked by a large measure of political freedom, and that has not also used something comparable to a free market to organize the bulk of economic activity.

Friedman emphasized the advantage of the unanimity that characterizes market transactions and decried the coercion required for governmental actions. But he was not an anarchist. He argued for limited government, not the chaos of the jungle. Here was his articulation of the rationale for government actions:

> There are clearly some matters with respect to which effective [market systems are] impossible. I cannot get the amount of national defense I want and you, a different amount. With respect

to such indivisible matters we can discuss, and argue, and vote. But having decided, we must conform. It is precisely the existence of such indivisible matters—protection of the individual and the nation from coercion are clearly the most basic—that prevents exclusive reliance on individual action through the market.

Other areas where Friedman believed government actions are necessary include: (1) developing and enforcing the legal system and property rights, (2) operating the monetary system, (3) controlling natural monopolies, and (4) dealing with neighborhood effects. The last one relates to dealing with externalities and will be considered further.

Friedman acknowledges neighborhood effects, which he defines as "effects on third parties for which it is not feasible to charge or recompense them." This is very close to our definition of an externality. Moreover, what Friedman calls "indivisible matters" in his discussion of national defense are similar to what are called public goods.

Friedman uses the example of national parks to illustrate his view of the appropriate treatment of neighborhood effects. Friedman argued that neighborhood effects "do not justify a national park, like Yellowstone National Park or the Grand Canyon." He explained as follows: "[If] the public wants this kind of an activity enough to pay for it, private enterprises will have every incentive to provide such parks. . . . I cannot myself conjure up any neighborhood effects or important monopoly effects that would justify governmental activity in this area."

Friedman's view is too narrow because he overlooks any inappropriable qualities of parks and similar environmental assets. By inappropriable, we mean activities that cannot be readily captured by private owners. Friedman argues in essence that a national park is really just an amusement park. In other words, the services of national parks only benefit visitors, and the benefits can be efficiently collected as toll charges by a private owner. The logic would be that if a mining company or developer found it more valuable, it should be able to buy Yellowstone, close it to public visits, and operate a huge open-pit mine to extract uranium.

Friedman's view of parks runs counter to modern environmental thinking on national parks and other treasures. Many places are precious to people: Venice to artists, Yellowstone National Park to naturalists, and New Mexico's Hermit's Peak to me and my family. The United Nations Educational, Scientific, and Cultural Organization (UNESCO) World Heritage Convention has a systematic process for listing major treasures. According to UNESCO, these sites are "among the priceless and irreplaceable assets, not only of each nation but of humanity as a whole." The list currently includes 1,092 sites around the world, including religious, ecological, and architectural monuments. Twenty-four are in the United States, including Yellowstone and the Grand Canyon (but not yet Hermit's Peak).

The guidelines for selection as a world heritage site are not just warm feelings or glossy pictures. Among the criteria for inclusion are that they contain superlative natural phenomena or areas of exceptional natural beauty and aesthetic importance, they represent a masterpiece of human creative genius, they bear a unique or at least exceptional testimony to a cultural tradition or civilization, or they represent an outstanding example of a type of building or architectural or technological ensemble. In their view (as well as that of most Americans), Yellowstone and the Grand Canyon meet these criteria.

Yellowstone as a Public Good

What is wrong with Friedman's approach to neighborhood effects? Using the language of economics, his approach ignores spillovers that are public goods rather than private goods. Recall that the key attributes of public goods include nonrivalry, meaning that the cost of extending the service to an additional person is zero, and nonexcludability, meaning that it is impossible to exclude people from enjoying them.

We used the example of lighthouses as public goods, and Yellowstone similarly has many public-goods qualities. According to the criteria of the World Heritage Center, here are its important features.

It has half the world's known geothermal features and the world's largest concentration of geysers. The park is one of the few remaining intact large ecosystems in the northern temperate zone of the earth. It is also a unique manifestation of wild ecosystems where rare and endangered species thrive. These benefits flow widely around the world and into the future, but they are unlikely to be reflected in the fees that people pay when visiting the park.

Because Yellowstone is managed as a public asset, it can preserve these unique features, its environmental quality can be assured, and people can enjoy it from afar. Measuring these values is extremely difficult, but they can plausibly be reckoned sufficiently large to keep the parks in public hands rather than turned over to private developers.

Friedman on Pollution Charges

In later writings with his wife, Rose Friedman, Milton Friedman looked more seriously at pollution. They recognized that pollution is sometimes dangerous, but regulations can be overly tight and badly designed. They wrote, "Most economists agree that a far better way to control pollution than the present method of specific regulation and supervision is to introduce market discipline by imposing effluent charges." The advantage of effluent charges, such as carbon taxes, is that they are transparent and operate efficiently. Note that, while they have kind words for a market approach, they do not endorse it.

The Moral Case for Free Markets

Free-market environmentalists make an interesting observation on the Green nature of innovation. The basic idea is that market forces will produce steady improvements in living standards. Technological change, in their view, is inherently Green because reducing environmental impacts is cost-beneficial for private industries. The Friedmans wrote in *Free to Choose*:

> If we look not at rhetoric but at reality, the air is in general far cleaner and the water safer today than one hundred years ago. The

air is cleaner and the water safer in the advanced countries of the world today than in the backward countries. Industrialization has raised new problems, but it has also provided the means to solve prior problems. The development of the automobile did add to one form of pollution—but it largely ended a far less attractive form.[5]

We can use the example of automobiles—so detested by many environmentalists—to illustrate the role of technological change in improving the environment. In the late nineteenth century, major cities were mired in filth from horse manure. At that time, New York had 100,000 horses operating as the major source of transportation. Alas, they also left behind 3 million pounds of manure and 10,000 gallons of urine a day, not to mention 25,000 horse carcasses to dispose of each year.

The invention and popularization of the automobile displaced horses as the major vehicle for urban transportation. Public health specialists of the time saw the car as the savior of health and welfare, and they were right. Cities sometimes banned horses from the streets, and horses are today primarily used as coaches for romantic interludes in New York's Central Park. The free-market point is that the new technology of automobiles was driven entirely by the lure of profits—for Henry Ford along with hundreds of other entrepreneurs. This clearly illustrates the free-market argument that growth is Green.

The history of lighting, discussed at length in chapter 10, illustrates how technology can improve the environment while raising living standards. For virtually all of human history, from open fires to oil lamps, energy efficiency improved at a snail's pace, perhaps 0.005% per year. Then, particularly with the introduction of electric lights, the energy efficiency of lighting improved dramatically, doubling every 12 years. With new technologies, humans not only spared whales but steadily decreased the pollution from fossil fuels. From 1970 to 2018, pollution from electricity for lighting per unit of light declined at more than 7% per year.

The list of environmental improvements can be extended indefinitely. Stepping back, two conclusions are evident. First, free-market

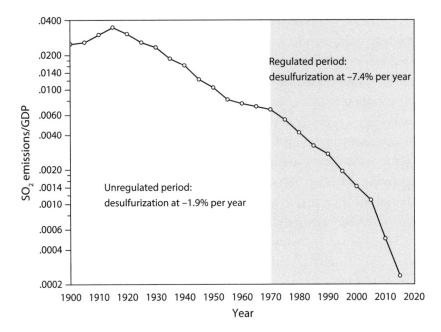

FIGURE 24-2. Sulfur emissions per unit output, 1900–2015

environmentalists are correct that private markets and public support for knowledge have been mighty engines of growth in living standards and, in many cases, environmental efficiency. Examples include horses, cars, lighting, and electronics.

However, unregulated markets did not get everything right. While free markets reduced whale oil for lighting, they also introduced electricity for lighting, and that electricity was generated by burning coal with its accompanying sulfur pollution. As described in the section on sulfur politics in chapter 14, sulfur dioxide emissions are one of the most damaging pollutants of the modern era. They were largely unregulated until 1970, after which emissions were increasingly tightly constrained.

The trend, measured as sulfur dioxide emissions per unit of gross domestic product (GDP), is shown in figure 24-2. In the early period, emissions were declining, largely because of efficiencies in electricity generation and the migration from coal to other energy sources. In the regulatory era starting in 1970, emissions declined even more rapidly. The rate of desulfurization went from −1.9% per year before 1970 to −7.4% per year after 1970. So, while free markets first dirtied the skies, they then helped clean the skies, but regulation helped even more.[6]

Of course, regulation was not without costs. The government collected comprehensive data on abatement costs from 1975 to 1994. Over that period, the costs of pollution abatement averaged 1.7% of GDP, with no trend in that ratio.

One point is clear: regulations did not wreck the economy. Moreover, our discussion of Green GDP showed that, when our economic accounts deal properly with the health benefits, environmental regulation has increased, not slowed, the growth of properly measured output.

The Goldilocks Rule of Regulation

How can we reconcile the insights of free-market advocates with the reality of regulatory history? We can apply the Goldilocks principle: Regulation should be neither too hot, nor too cold, but just right. In other words, it is necessary to find the appropriate balance between no regulation and draconian regulation.

Figure 24-3 makes the point using the example of a carbon tax to show the impacts of regulation on *true income,* or income corrected for the harmful effects of externalities. *Measured income* is conventional GDP, which includes the cost of pollution abatement but not the benefits. *True income* includes both abatement costs and damages. Additionally, the little bubbles show the levels of a carbon tax that maximize each of the two measures of income.

Measured income, such as standard GDP, which excludes averted damages, is maximized at zero tax and zero abatement. True income has its maximum at the optimal tax rate of $40 per ton of carbon. Hence, proper measurement shows that true income is maximized when appropriate environmental policies are taken, at the Goldilocks level.

Thus, the useful message of free-market environmentalists is this: Do not overdo your Green enthusiasm. Regulations can be too hot as well as too cold. When they are too hot, they will choke the spirit of

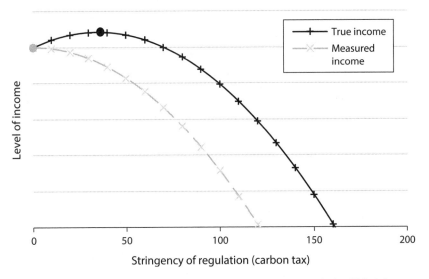

FIGURE 24-3. The Goldilocks principle on the environment is that controls should find the happy middle between doing nothing and overdoing. Note that conventionally measured income is maximized with no abatement, while true income is maximized at the Goldilocks point where marginal benefits and marginal damages are equal.

enterprise. It would be better to allow the development of polluting automobiles than to ban automobiles and hire an army of sweepers to clean up the piles of horse manure. But it is better still to allow the innovative spirit to thrive by having a light regulatory footprint and, as Friedman insisted, make maximal use of market instruments.

The Chicago School on Regulation

Market advocates have a double-barreled attack on environmental activism. The first barrel is skepticism about the damages from environmental degradation. Skeptics might see the trends in climate change but question whether the impacts are as dire as scientists claim. Some might even argue that rising CO_2 levels will be a benefit because CO_2 is a fertilizer that will raise agricultural production. The skeptical view on damage impacts has not been supported by recent research, but it needs careful attention.

A second barrel in the attack on Green is emphasizing *government failures*. A government failure can arise if the government selects a policy, such as subsidizing energy, that leads to an inefficient outcome. Government failures also arise when interest groups successfully lobby for interventions that promote their interests rather than the public interest.

The Chicago School's approach to regulation has been particularly influential for economic regulations, such as rules that limit competition in industries like airlines, trucking, and electricity generation. For example, numerous economic studies have shown that economic regulation often keeps prices high. For many years trucking companies and airlines had to get permission before lowering prices or entering new markets.

Do government failures apply to environmental regulation as well? The answer is yes, but in a different way. In their review of government failures in environmental regulation, David Anthoff and Robert Hahn identified several in which regulations could be substantially improved.[7] Here are some key examples:

- *Loss of revenues from limiting emissions.* Governments generally limit pollution by issuing permits for free to incumbents in an industry. Two important examples were sulfur dioxide in the United States and CO_2 in Europe. While offering free permits to industry may reduce political resistance, it loses precious revenues and makes the tax system less efficient, as explained in the chapter on Green taxation. This approach also tends to lock in existing companies and technologies.

- *Poor analysis.* The gold standard for analysis of environmental regulations is cost-benefit analysis. Such analyses ensure that the incremental costs are balanced against the incremental damages, which is necessary to meet the Goldilocks principle. Limitations on cost-benefit analysis are sometimes embedded in law, but more often they arise from a reluctance of administrators to confront both sides squarely. The result is that environmental regulations are a hodgepodge of too strict and too lax.

- *Ignoring scarcity of public resources.* A third issue arises because many public resources are treated as free when in fact they are scarce. Important examples here include fossil underground water, roads, and airports, as well as the more obvious ones of clear air and clean water. Pricing congestion on roads is an important step in both limiting the huge waste of people's time in traffic jams and raising funds to repair the nation's decaying infrastructure.

- *Global public goods.* Another pervasive failure comes from global public goods, like climate change. Here, the temptation of individual countries to free ride on the efforts of other countries produces too low a level of global abatement.

These are among the many examples of why environmental regulation is an imperfect instrument for meeting Green goals. The implication is not that we should abandon the effort. Rather, it emphasizes the need for hardheaded analysis of the goals and careful attention to the means of attaining Green goals.

The Free-Market Case for a Carbon Tax

There is no better example of a light regulatory footprint than in policies to slow climate change. The analysis of global Green discussed the threat of climate change. How might a free-market environmentalist think about climate-change policies? The following are the musings of a hypothetical environmentally oriented conservative on this question.[8]

"As a conservative, I desire a political and economic system that is efficient, equitable, and has maximum individual freedom. However, I also desire to leave a better world for my children and grandchildren. I am no defender of big oil or corporate irresponsibility, and I don't think that anyone should be allowed to despoil the earth at other people's expense. I might think that this was well expressed by the conservative U.S. president Ronald Reagan:

> If we've learned any lessons during the past few decades, perhaps the most important is that preservation of our environment is not a partisan challenge; it's common sense. Our physical health, our social happiness, and our economic well-being will be sustained

only by all of us working in partnership as thoughtful, effective stewards of our natural resources.[9]

"So, wearing my conservative hat, I begin by reading the scientific analyses. After reading the science with an open mind, I conclude that the evidence behind climate-change science is convincing. There are lots of ifs, buts, and qualifications. But the idea that armies of scientists around the world are conspiring to perpetrate a giant hoax, or that climate change is a Chinese conspiracy to boost their manufacturing, is just silly.

"I then study the literature on impacts. The evidence here is much murkier because we are projecting uncertain climate projections on rapidly changing future societies. But I find the projections very unsettling. I might have a fine beach house and read that it is likely to wash into the sea. I read about the forced migration of millions of people and wonder whether they will spill over to my town, state, and country. I read that climate change is destroying many of the natural wonders of the world that I would hope to visit with my grandchildren.

"Finally, I turn to the policy-makers. How about turning it over to the market? I quickly realize that we definitely cannot rely on a pure 'free-market' solution, which involves no restraint on carbon emissions. Some kind of governmental intervention is necessary to slow global warming.

"Environmental activists appear to favor a 'cap-and-trade' approach, which sets up an allocation of allowances to emit CO_2 and gives them away to deserving parties. Activists are apparently imposing regulations on automobiles, power plants, appliances, and light bulbs. One of my favorite conservative talk-show hosts denounces this as 'light-bulb socialism,' and that sounded funny and right. The current approach favored by many environmentalists and governments has a heavy regulatory footprint, and one that is not all that effective.

"What do the economists say here? Naturally, begin with my hero, Milton Friedman. He favors effluent charges. Many economists are advocates of something called a 'carbon tax.' This would impose a tax on emissions of CO_2 and other greenhouse gases. It would accomplish the goal of raising the price of CO_2 emissions to cover their social costs.

"What do conservative economists think? I look at the writings of conservative economists Martin Feldstein (chief economist to Ronald Reagan), Michael Boskin (chief economist to George H. W. Bush), Greg Mankiw (chief economist to George W. Bush), Kevin Hassett (chair of the Council of Economic Advisers under President Trump), Arthur Laffer (of Laffer curve fame), and Gary Becker (Nobel Prize–winning Chicago-school economist). They all favor a carbon tax as the most efficient approach to slowing global warming.

"Their point is that those who burn fossil fuels are enjoying an economic subsidy—in effect, they are grazing on the global commons and not paying for what they eat. A carbon tax would improve economic efficiency because it would correct for the implicit subsidy on the use of carbon fuels.

"I conclude that carbon taxes are an ideal policy for true free-market conservatives who care about preserving our beautiful planet but want to do so with market-based incentives and with minimal government intrusion. The carbon tax and similar market-based policies are approaches on which proponents of Green and free-market environmentalism agree."

Thus spake a free-market environmentalist.

Summary on Green Skeptics

How can we summarize the view of Green skeptics? To begin with, they have many different views. Some are just advocating for their profits and private interests—perhaps because they own coal companies or work in polluting industries. While we can recognize their positions, we should not confuse private profits with the public interest.

Additionally, we need to recognize the validity in the free-market philosophy where it applies. Economic history shows that innovation and technological change have generally been Green because new technologies use less energy, and less energy generally means less pollution.

For most of the period since the industrial revolution, there was little or no regulation on pollution. Increasingly, since 1970, governments have put controls on most major pollutants (except for greenhouse gases). For the United States, the cost of pollution control has been slightly below 2% of GDP. Careful analysis indicates that benefits outweighed the costs, so true income and growth were increased, not decreased, by controls.[10]

So even the most fervent environmentalist should take seriously the arguments of free-market environmentalists. Effective environmental policy requires sound science, careful balancing of costs and benefits, and the design of effective mechanisms for implementing policies. The history of centrally planned economies demonstrates the dead hand of overzealous central control while the failure of climate-change policies shows the dangers at the opposite pole of inaction. Environmentalism and market orientation intersect with efficient policies such as carbon taxes, auction of public resources, and minimizing the role of command-and-control mandates.

25

A Tour of the Spirit of Green

So ends the tour of Planet Green. The voyage has shown how we humans interact with ourselves, with other species, and with natural ecosystems. The interaction has produced astonishing economic progress, but it has also been accompanied by undesired collisions and contagions along the way.

In an earlier era, when the first European settlers arrived in my home state of Connecticut, the major problems were coping with the natural elements. Life was filled with clearing trees for agricultural land, staying warm in the brutal winters, and battling horrible diseases. Neighbors were necessary for protection.

As our continent and world have filled up with people, factories, roads, and pollution, our neighbors are harming us as well as protecting us. Brown is crowding out Green. We see it in pollution, waste, congestion, litter, depletion of species, overfishing, and, most ominously, climate change.

These are serious issues that may arise in a crowded world, and they may spin out of control if ignored. Our best estimate is that the gains brought by technological advances and international trade have outweighed the damages of pollution and other externalities. But there is no iron law of politics or the market to ensure that the upward trend will continue.

The book rests its analysis and ethical perspective on the *goals of a well-managed society*—one designed to advance the well-being of its members. It rests on four pillars. These include laws to define property rights and contracts so that people can interact fairly and efficiently; effective markets to engage in exchange of private goods; laws, regulations, expenditure, and taxes to correct important externalities and provide public goods; and corrective taxation and expenditure to help ensure appropriate equity in the distribution of economic welfare.

To cope with the *undesirable side effects of growth*, we must recognize the proper roles of market and government. The market cannot solve all our social problems any more than can the government. The market by itself cannot effectively curb climate change; the government by itself cannot effectively allocate bread or oil. Finding the right mixture of market and government is one of the most vexing issues of economic and environmental policies. Each plays a central role in maintaining the balance between improving living standards and controlling pollution.

A central theme of the Green discussion in this book is the role of *efficiency*. Efficiency is the staple of economists and denotes the most effective use of a society's resources in satisfying people's wants and needs. While we often laud the effectiveness of properly functioning markets (for example in providing life-saving vaccines), we also recognize situations in which markets fail, such as in the presence of negative externalities like pollution or contagious diseases. Activities with negative externalities lead to unintended spillovers in which those who benefit do not compensate those who are harmed.

For *negative externalities*, the presumption is that the unregulated market will misallocate resources, producing too much Brown and too little Green. In some areas, the externalities are relatively small, and we choose to tolerate the results. For example, traffic congestion wastes billions of hours a year. Economists have devised ingenious schemes to put a price on congestion, but most countries have decided to groan and bear it rather than pricing it. In other areas, such as deadly air pollution from burning coal, countries have taken steps to curtail the worst dangers.

One of the central principles of Green thinking is its emphasis on *sustainability*. A sustainable path for the economy is one that allows every future generation the option of being as well off as its predecessors. But we do not insist on being as well off in every dimension—for every good and service and enjoyment. A central approach of economics is to emphasize consumption substitutability, which allows consumers to meet their needs by substituting goods with declining prices for those with rising scarcity. In the context of sustainability, this means we should be primarily concerned about people's living standards in food, shelter, health care, and the rest rather than with how they are produced. As an example, it matters less whether goods are recycled than whether they will quickly degrade into innocuous substances at the end of their useful lifetimes. While there are exceptions, such as unique and irreplaceable assets such as Yellowstone, resources are generally valued for what they do, not what they are.

The concept of sustainability finds an important application in Green national accounts. The standard economic accounts (such as for gross domestic product, or GDP) largely omit the impacts of externalities like health damages from pollution. Including the economic effects of such externalities can make a substantial difference to the *level* of output. An estimate using existing research suggests that correcting for omissions would subtract on the order of 10% from output for the United States.

However, and paradoxically, correcting for externalities will tend to raise the *growth rate* of true output, at least for the last half century for the United States. This is because the emissions of most pollutants have been declining relative to the overall economy. So those who complain about the impacts of environmental regulations on economic growth are really complaining about measurement, not actual impacts.

Green policies are most often associated with combating pollution or congestion. However, infectious diseases similarly display the syndrome of harmful externalities of economic activity and globalization. They require different tools, such as government-directed treatments and vaccines, but are still examples of harmful spillovers that must be corrected.

Additionally, pandemics are part of a deadly syndrome of fat-tailed catastrophes—phenomena where low-probability and high-consequence events may take place. These tail events are especially challenging exactly because they are rare. We cannot accurately predict their frequency or severity, which in turns makes it difficult to recognize them when they emerge, and equally difficult to prepare for them in advance.

A related point is to recognize that correcting externalities is costly. At the very least, such corrections require the scarce time of governments and company managers who have competing concerns. From an economic point of view, most interventions take place through the regulatory process of the command-and-control variety ("Do this but don't do that."). Regulations involve necessary costs of compliance (to install pollution control equipment, for example) but also have excess costs because they are difficult to design with perfect or even reasonable efficiency. The excess cost of regulation reinforces the point that governments must choose which Brown problems to control and which to leave alone. Just as important is that policies should use the most efficient tools available.

A hopeful new trend in the design of environmental policies is the use of *market mechanisms*, particularly pollution taxes, to control externalities. These have proven extremely effective in reducing conventional air pollution. such as sulfur dioxide. Many economists believe that the single best tool for slowing climate change is to use high carbon prices, such as through imposing carbon taxes, as a way of restraining carbon dioxide (CO_2) emissions as well as providing incentives for low-carbon innovation.

The next set of issues involves the perverse effect of *defective decision-making* (the subject of behavioral economics). Perhaps the most prevalent mistake people make is to ignore life-cycle costs and focus on first cost. We see this most vividly in decisions about energy use (too much fuel use and too little energy-saving first-cost capital). First-cost bias is related to the issue of too-high discount rates, which similarly place too much emphasis on early costs and tend to ignore distant costs. Many behavioral anomalies (particularly too-high discount rates) have adverse environmental consequences because

Green projects are often those that involve up-front capital costs (which are overweighted) and future environmental benefits (which are underweighted). Behavioral issues require different approaches from externalities—sometimes better information, sometimes regulations, and sometimes new technologies.

We have also reviewed *the applications of Green philosophy* to important areas such as politics, innovation, corporate responsibility, and investment. In each of these areas, a central dilemma arises around the trade-off between benefits to the decision-maker and broader societal goals. For all areas, a key reminder is that decision-makers need to avoid short termism and take a broad view of what improves long-run outcomes, whether profits, returns, or social welfare. Additionally, each institution has its own expertise, and therein lies an important responsibility. Corporations, universities, investors, and governments need to apply their specialized knowledge, and in the case of firms, to ensure that they provide clear and unbiased information about the safety of their products and processes.

One of the most important concepts in undertaking Green actions is the principle of *no regrets*. When our actions cause harmful spillovers, small reductions in our externality footprint have very small impacts on our own welfare but large reductions in harm to others. In other words, by taking small steps we can reduce our spillovers, perhaps substantially, without experiencing any regrets because there are almost no impacts on us. This principle can be usefully applied to areas such as reducing our carbon, pollution, and congestion footprints.

One area where there is no dilemma in the application of Green principles is *environmental taxes*. Green taxes use fiscal instruments to internalize negative externalities like pollution. Environmental taxes are one of the most promising innovations of recent years. Green taxes are the holy grail of public policy. They have a trinity of traits: they pay for valuable public services, they meet environmental objectives efficiently, and they are nondistortionary. The areas with the largest potential are carbon taxes and gasoline taxes, with sin taxes on alcohol, tobacco, firearms, and gambling being closely related. Here is a way of thinking about this approach: "Tax bads, not goods." This little adage is simple, intuitive, and correct.

As we examine many of the challenges that today's economies face, we find that solutions will often require technological changes. One historical example was how the mountains of horse manure produced when transporting people and goods were cleaned up not by sweepers, but by the automobile. Most recently, emissions of sulfur dioxide declined sharply as a result of economic incentives as well as improved institutional and technical innovations. A transition to a low-carbon world will depend upon technical advances to replace our fossil-dependent technologies.

The discussion above has emphasized the strong headwinds faced by Green innovative activities. *Innovation for environmental products and services has a special challenge in what can be called a double externality.* Green production is not only underpriced but the private returns to innovation are below the public returns. So the first externality is a gap between the private cost and the social cost of pollution. However, there is a further gap between the social and the private returns on innovation. Putting these two together can virtually erase incentives of profit-oriented firms to pursue environmentally friendly innovation. Correcting the pollution externality is an important step and will correct the pollution-externality gap, but that still leaves the innovation gap uncorrected. Helping address the double externality in Green innovation is one of the most urgent reasons to internalize pollution externalities.

A final frontier is global environmental issues, or *global public goods*. While many global threats exist, climate change is the ultimate Green challenge. The chapters on global Green contain four key findings. First, global citizens need to understand and accept the gravity of the impacts of global warming on the human and natural world. People should be alert to the trumped-up claims of contrarians who find some negative results or list reasons to wait for decades to take the appropriate steps.

Second, nations must establish policies that raise the price of CO_2 and other greenhouse-gas emissions. While experts recognize the importance of carbon pricing, virtually no progress has been made on a global scale. We need to ensure that actions are global. While politics may be local—and the opposition to strong steps to slow warming comes from nationalistic attitudes—slowing climate change requires coordinated global action.

The best hope for effective coordination on an international level is a climate compact, a coalition of nations who commit to strong steps to reduce emissions along with mechanisms to penalize countries that do not participate.

———

I am often asked if I am discouraged about the slow progress in achieving our Green objectives. The efforts of a progressive administration were then set back by the pro-pollution and corrupt members of the next administration. The Obama administration struggled to implement a strong climate-change policy, but the Trump administration uprooted all progress and declared that climate change is a Chinese invention to advance its manufacturing industries. History turned the page again with the Biden administration, faced by environmental, economic, and public-health crises on many fronts.

It is easy to become cynical when today's national leaders are so ignorant and venal. What might be the inconsequential behavior of individuals leads to Earth-threatening consequences for societies. The historian Barbara Tuchman aptly described this syndrome:[1]

> Wooden-headedness, the source of self-deception, is a factor that plays a remarkably large role in government. It consists in assessing a situation in terms of preconceived fixed notions while ignoring or rejecting any contrary signs. It is acting according to wish while not allowing oneself to be deflected by the facts. It is epitomized in a historian's statement about Philip II of Spain, the surpassing wooden-head of all sovereigns: "No experience of the failure of his policy could shake his belief in its essential excellence."

Denial of climate change and provoking trade wars today, like denial of the dangers of smoking and launching a war in Iraq in earlier eras, are woodenheadedness that imperils our planet and its inhabitants.

We might become either optimistic or pessimistic about our ability to cope with threats to our Green future. On the one hand, it is true that we are moving into uncharted waters, depleting many resources while altering others in an irreversible manner and gambling with our Earth systems and future climate. Humans are quarrelsome and have devised weapons that are awesomely effective at avenging their disputes.

At the same time, our scientific knowledge and capabilities are much more powerful than they were in earlier years. And the growing Spirit of Green provides both the scientific basis and the popular support for policies that can overcome the deadly spillovers of economic growth.

What will prevail in the race between our tendency to quarrel and pollute and our power to reason and compute? The jury is out. But if we can face the future with intellectual honesty and farsightedness, we have the tools and resources to realize the dream of a Green Earth.

Chapter 2. Green History

1. The quotes from Gifford Pinchot are from his book *A Primer of Forestry*, U.S. Department of Agriculture, Division of Forestry, Bulletin No. 24, vol. 2 (Washington, DC: Government Printing Office, 1903–1905).

2. John Muir, *A Thousand-Mile Walk to the Gulf* (Boston: Houghton Mifflin, 1916), http://vault.sierraclub.org/john_muir_exhibit/writings/a_thousand_mile _walk_to_the_gulf. The quotation on alligators is from John Muir, *John of the Mountains: The Unpublished Journals of John Muir*, ed. Linnie Marsh Wolfe (Madison: University of Wisconsin Press, 1979); John Muir and Michael P. Branch, *John Muir's Last Journey: South to the Amazon and East to Africa: Unpublished Journals and Selected Correspondence*, vol. 52 (Washington, DC: Island Press/Shearwater Books, 2001), https://catalog.hathitrust.org/Record/004179556.

3. Two important works on biocentrism and deep ecology are Bill Devall and George Sessions, *Deep Ecology* (Salt Lake City: G. M. Smith, 1985) and Paul Taylor, *Respect for Nature: A Theory of Environmental Ethics*, Studies in Moral, Political, and Legal Philosophy (Princeton, NJ: Princeton University Press, 1986).

4. One of the leading proponents of biocentric philosophy was Paul Taylor in *Respect for Nature*. This is also called *deep ecology* after Arne Næss, "The Shallow and the Deep, Long-Range Ecology Movement. A Summary," *Inquiry* 16, no. 1–4 (January 1, 1973): 95–100, doi:10.1080/00201747308601682. The history of including animals in social preferences goes back to the founding of utilitarian philosophy as advocated by Jeremy Bentham and John Stuart Mill.

5. Taylor, *Respect for Nature*, 13.

6. Muir, *A Thousand-Mile Walk to the Gulf*, 98, 139.

7. Garrett Hardin, "The Tragedy of the Commons," *Science* 162, no. 3859 (December 13, 1968): 1243–48, doi:10.1126/science.162.3859.1243.

8. Ibid., 1244.

9. Ibid., 1248.

10. Ibid., 1244.

11. Rachel Carson, "Undersea," *The Atlantic Monthly*, September 1937, 322.

12. Rachel Carson, *Silent Spring*, ed. Lois Darling and Louis Darling (Boston: Houghton Mifflin, 1962), https://archive.org/stream/fp_Silent_Spring-Rachel _Carson-1962/Silent_Spring-Rachel_Carson-1962_djvu.txt.

13. Carson, "Undersea," 266.

14. An interesting history of Rachel Carson's role in spurring the environmental policy of the Kennedy administration is contained in Douglas Brinkley, "Rachel Carson and JFK, an Environmental Tag Team," *Audubon*, May/June 2012.

15. Daniel C. Esty, *A Better Planet: 40 Big Ideas for a Sustainable Future* (New Haven, CT: Yale University Press, 2019).

16. Esty, *Better Planet*, essay 7.

17. John Maynard Keynes, *The General Theory of Employment, Interest and Money* (New York: Harcourt, Brace, 1936), 383–84.

Chapter 3. Principles of a Green Society

1. The economic conception of a good society is found in many writings. A good example that relates closely to the present study is Francis Bator, "The Anatomy of Market Failure," *Quarterly Journal of Economics* 72, no. 3 (August 1958): 351–79.

2. The discussion of a well-ordered society is from several works by John Rawls, including *A Theory of Justice* (Cambridge, MA: Harvard University Press, 1965); "Justice as Fairness: Political Not Metaphysical," *Philosophy and Public Affairs* 14, no. 3 (1985): 223–51; "Reply to Alexander and Musgrave," *Quarterly Journal of Economics* 88, no. 4 (1974): 633–55, doi:10.2307/1881827.

Chapter 4. Green Efficiency

1. The definition of "efficiency" and the discussion of the invisible-hand principle draw heavily on Paul Samuelson and William Nordhaus, *Economics*, 19th ed. (Boston: McGraw-Hill Irwin, 2010).

2. Adam Smith, *An Inquiry into the Nature and Causes of the Wealth of Nations*, vol. 2 (London: W. Strahan and T. Cadell, 1776), 35.

3. This foundational work began as *Wealth and Welfare* (London: Macmillan, 1912), which was revised as *The Economics of Welfare* (London: Macmillan, 1920). The book went through four editions, and the last edition from 1932 can be found online at https://www.econlib.org/library/NPDBooks/Pigou/pgEW .html.

A fine biography that describes Pigou's life and times can be found in Ian Kumekawa, *The First Serious Optimist: A. C. Pigou and the Birth of Welfare Economics* (Princeton, NJ: Princeton University Press, 2017).

4. The quotations from Pigou in this section are from his *Economics of Welfare*, part 2, chap. 9, sect. 3, slightly abridged for clarity.

5. Francis M. Bator, "The Simple Analytics of Welfare Maximization," *American Economic Review* 47, no. 1 (1957): 22–59.

6. The distinction between public and private goods, the treatment of networks, and the discussion of lighthouses are drawn from Samuelson and Nordhaus, *Economics*.

7. The study on the economic impact of job losses is from Steven J. Davis and Till von Wachter, "Recessions and the Costs of Job Loss," *Brookings Papers on Economic Activity*, no. 2 (2011): pp. 1–73.

Chapter 5. Regulating Externalities

1. If you have studied economics, you have encountered the term *marginal* many times. Here is a simple explanation. You might grow tomatoes in your garden, and once they are planted, your main cost is your time. Suppose that when you spend ten hours you get ten tomatoes, with eleven hours you get twelve tomatoes, with twelve hours you get thirteen tomatoes, and with thirteen hours you still get thirteen tomatoes because everything useful has been done. The "marginal tomato output" (MTO) is defined as the increment of tomatoes per additional hour. So the MTO of going from ten to eleven hours is two tomatoes, from eleven to twelve hours is one tomato, and from twelve to thirteen hours is zero tomatoes. If you value your time as one-half tomato per hour, you will work twelve hours because the last hour has an MTO of one tomato, and the cost of that is one-half tomato of time cost. It would not make sense to work thirteen hours because the marginal output is zero. The basic idea of economics is that the marginal benefit (here tomatoes) should be at least the marginal cost (here work). For pollution control, the idea is that the marginal benefit (improved health) should be at least as large as the marginal cost (in labor, capital, and other inputs).

2. Table 5-1 shows large steps up to the level close to the optimal pollution. At that point we have shown the results at a finer grain of resolution. Net benefits hardly change at the maximum net benefits in the last column. This is similar to the tiny loss of altitude at the top of a gentle hill if you move a few feet in any direction.

Chapter 6. Green Federalism

1. George W. Downs and David M. Rocke, "Conflict, Agency, and Gambling for Resurrection: The Principal-Agent Problem Goes to War," *American Journal of Political Science* 38, no. 2 (1994): 362–80, doi:10.2307/2111408.

Chapter 7. Green Fairness

1. On fairness and justice, see Michael J. Sandel, *Liberalism and the Limits of Justice*, 2nd ed. (Cambridge: Cambridge University Press, 1998) and Amartya Sen, *The Idea of Justice* (Cambridge, MA: Belknap Press of Harvard University Press, 2009).

2. James J. Heckman, *Giving Kids a Fair Chance*, Boston Review Books (Cambridge, MA: MIT Press, 2013).

3. The study on the incidence of gasoline taxes is from Antonio M. Bento, Lawrence H. Goulder, Emeric Henry, Mark R. Jacobsen, and Roger H. von Haefen, "Distributional and Efficiency Impacts of Gasoline Taxes: An Econometrically Based Multi-market Study," *American Economic Review* 95, no. 2 (2005): 282–87, doi:10.1257/000282805774670536.

4. U.S. Environmental Protection Agency, *The Benefits and Costs of the Clean Air Act, 1970 to 1990, Prepared for U.S. Congress by U.S. Environmental Protection Agency* (October 1997), https://www.epa.gov/sites/production/files/2017-09 /documents/ee-0295_all.pdf.

5. See Michael Ash and T. Robert Fetter, "Who Lives on the Wrong Side of the Environmental Tracks? Evidence from the EPA's Risk-Screening Environmental Indicators Model," *Social Science Quarterly* 85, no. 2 (2004): 441–62.

6. To read more and see some of the selfies, go to https://en.wikipedia.org /wiki/Monkey_selfie_copyright_dispute.

7. On pain in crabs, see Robert W. Elwood and Mirjam Appel, "Pain Experience in Hermit Crabs," *Animal Behaviour* 77, no. 5 (May 1, 2009): 1243–46, doi:10.1016/j.anbehav.2009.01.028.

8. Dwight D. Eisenhower, "Chance for Peace" (speech), April 16, 1953, Miller Center, University of Virginia, transcript, https://millercenter.org/the-presidency /presidential-speeches/april-16-1953-chance-peace.

Chapter 8. Green Economics and Concepts of Sustainability

1. A survey of Green economics that covers many of the key ideas is in Miriam Kennet and Volker Heinemann, "Green Economics: Setting the Scene: Aims, Context, and Philosophical Underpinning of the Distinctive New Solutions Offered by Green Economics," *International Journal of Green Economics* 1, no. 1–2 (2006): 68–102, doi:10.1504/IJGE.2006.009338.

2. Michael Jacobs, *The Green Economy: Environment, Sustainable Development and the Politics of the Future* (Vancouver: UBC Press, 1993).

3. Ibid., 72.

4. The quotes on sustainable development are from the World Commission on Environment and Development, *Our Common Future* (Oxford: Oxford University Press, 1987), 2, 43.

5. The quotations and the discussion regarding the views of Robert Solow are from "An Almost Practical Step toward Sustainability: An Invited Lecture on the Occasion of the Fortieth Anniversary of Resources for the Future" (lecture, Washington, DC, October 8, 1992).

6. The discussion here draws on William Nordhaus, "Is Growth Sustainable?," in *Economic Growth and the Structure of Long-Term Development: Proceedings of the IEA Conference Held in Varenna, Italy*, ed. Luigi L. Pasinetti and Robert M. Solow (Houndmills, Basingstoke: Macmillan, 1994), pp. 29–45.

7. A good place to start for long-term forecasts is the Congressional Budget Office (CBO) at www.cbo.gov. The CBO did not project a major decline in long-run growth in its postpandemic outlooks. Similarly, the Federal Reserve projections suggest a "longer-run" growth in real GDP that is 0.1% per year slower after the pandemic than the projection before the pandemic.

8. Peter Christensen, Ken Gillingham, and William Nordhaus, "Uncertainty in Forecasts of Long-Run Economic Growth," *Proceedings of the National Academy*

of Sciences of the United States of America 115, no. 21 (May 22, 2018): 5409–14, doi:10.1073/pnas.1713628115.

9. There is an important qualification here that, as noted earlier in this section, standard measures omit corrections for externalities such as the health impacts of pollution. The next chapter examines the potential size of this correction.

10. Solow, "Almost Practical Step toward Sustainability."

11. Jeffrey Sachs, "Sustainable Development: Goals for a New Era" (lecture, Pontifical Academy of Sciences and the Pontifical Academy of Social Sciences, Vatican, Vatican City, Rome, May 2014).

Chapter 9. Green National Accounting

1. The inspirational quote from the young radical at the beginning of this chapter led to an early work on Green national accounts, William Nordhaus and James Tobin, "Is Growth Obsolete?," in vol. 5, *Economic Research: Retrospect and Prospect*, NBER Book Chapter Series, no. c7620, ed. William Nordhaus and James Tobin (Cambridge, MA: National Bureau of Economic Research, 1972), pp. 509–564.

2. Paul Samuelson and William Nordhaus, *Economics*, 19th ed. (Boston: McGraw-Hill Irwin, 2010), with a few changes for context.

3. The theory of national accounting with externalities was developed by Martin L. Weitzman, "On the Welfare Significance of National Product in a Dynamic Economy," *Quarterly Journal of Economics* 90, no. 1 (1976): 156–62, doi:10.2307/1886092.

4. National Research Council, *Nature's Numbers: Expanding the National Economic Accounts to Include the Environment* (Washington, DC: National Academies Press, 1999).

5. See the table source note for the sources and methods.

6. Estimates of the impact of correcting for subsoil assets are predominantly from Bureau of Economic Analysis, *Survey of Current Business*, April 1994, and discussed in National Research Council, *Nature's Numbers*. The U.S. work on resource accounting was pioneered by Steve Landefeld, the former director of the U.S. Bureau of Economic Analysis, who was instrumental in modernizing the U.S. national economic accounts.

7. Estimates of the cost of air pollution are from Nicholas Z. Muller, Robert Mendelsohn, and William Nordhaus, "Environmental Accounting for Pollution in the United States Economy," *American Economic Review* 101, no. 5 (2011): 1649–75, doi:10.1257/aer.101.5.1649. Estimates have since been updated by Nicholas Z. Muller, "Boosting GDP Growth by Accounting for the Environment," *Science* 345, no. 6199 (2014): 873–74, doi:10.2307/24917200.

Chapter 10. The Lure of Exo-civilizations

1. Estimates of early living standards are from Angus Maddison, *Contours of the World Economy, 1–2030 A.D.: Essays in Macro-economic History* (Oxford: Oxford University Press, 2007). Recent data are from the International Monetary Fund. Estimates for the earliest times are from Brad de Long, "Estimates of World GDP,

One Million B.C.–Present," *DeLong: Long Form* (blog), 1998, https://delong
.typepad.com/print/20061012_LRWGDP.pdf and use estimates of subsistence
output for the earliest period.

2. The history of lighting is based on my article "Do Real-Output and Real-Wage
Measures Capture Reality? The History of Lighting Suggests Not," NBER Book
Chapter Series, no. c6064 (Cambridge, MA: National Bureau of Economic Research,
1996). I have updated the estimates from that study for the present chapter.

3. Louis Stotz, *History of the Gas Industry* (New York: Press of Stettiner Bross,
1938), 6.

4. Cited at *Elon Musk News* (blog), https://elonmusknews.org/blog/elon
-musk-spacex-mars-quotes.

5. Ross Andersen, "Exodus," *Aeon*, September 30, 2014, https://aeon.co/essays
/elon-musk-puts-his-case-for-a-multi-planet-civilisation.

6. Adam Morton, *Should We Colonize Other Planets?* (Cambridge: Polity Press,
2018); Sydney Do, Andrew Owens, Koki Ho, Samuel Schreiner, and Olivier de
Weck, "An independent Assessment of the Technical Feasibility of the Mars One
Mission Plan—Updated Analysis," *Acta Astronautica* 120 (2016): 192–228.

7. There is very little serious research on Biosphere 2. A key article is from
Joel E. Cohen and David Tilman, "Biosphere 2 and Biodiversity: The Lessons
So Far," *Science* 274, no. 5290 (1996): 1150–51, doi:10.1126/science.274.5290.1150.
For an enthusiastic and largely overoptimistic account, see John Allen and Mark
Nelson, "Overview and Design: Biospherics and Biosphere 2, Mission One (1991–
1993)," *Ecological Engineering* 13 (1999): 15–29.

8. Most estimates in the accounting are from Cohen and Tilman, "Biosphere
2 and Biodiversity," 1150–51.

Chapter 11. Pandemics and Other Societal Catastrophes

1. The history and epidemiology of pandemics are described in an outstanding
book by my Yale colleague, Nicholas Christakis, *Apollo's Arrow* (New York: Little,
Brown Spark, 2020).

2. Data are from Pasquale Cirillo and Nassim Nicholas Taleb, "Tail Risk of
Contagious Diseases," *Nature Physics* 16 (2020): 606–13, doi:10.1038/s41567-020
-0921-x.

3. See the report from the Centers for Disease Control and Prevention, *Coronavirus Disease 2019 (COVID-19)*, https://www.cdc.gov/coronavirus/2019-ncov
/cases-updates/commercial-lab-surveys.html.

4. CNN, "Fauci Says the WHO's Comment on Asymptomatic Spread Is Wrong,"
https://www.cnn.com/2020/06/09/health/asymptomatic-presymptomatic
-coronavirus-spread-explained-wellness/index.html.

5. Office of Management and Budget, *Budget of the U.S. Government, Fiscal
Year 2021*, https://www.govinfo.gov/app/collection/budget/2021.

6. J Joseph A. Schumpeter, "The Crisis of the Tax State" in *International Economic Papers* 4, eds. A. T. Peacock, R. Turvey, W. F. Stolper, and E. Henderson

(London and New York: Macmillan, 1954): 5–38 [translation of "Die Krise des Steuerstaates," *Zeitfragen aus dem Gebiet der Sociologie* 4 (1918): 1–71].

7. John Witt, *The Legal Structure of Public Health* (New Haven, CT: Yale University Press, 2020).

8. U.S. Homeland Security Council, *National Strategy for Pandemic Influenza: Implementation Plan* (May 2006): 4.

9. Bob Woodward, *Rage* (New York: Simon and Schuster, 2020).

Chapter 12. Behavioralism as the Enemy of the Green

1. George Loewenstein and Richard H. Thaler, "Anomalies: Intertemporal Choice," *Journal of Economic Perspectives* 3, no. 4 (1989): 181–93, doi:10.1257 /jep.3.4.181; with small edits for context.

2. David Laibson, "Hyperbolic Discount Functions, Undersaving, and Savings Policy" NBER Working Paper 5635 (Cambridge, MA: National Bureau of Economic Research, June 1996), https://doi.org/10.3386/w5635.

3. First-cost bias goes by many names. It is also called the energy-efficiency gap and the energy paradox. For a skeptical view, see Hunt Allcott and Michael Greenstone, "Is There an Energy Efficiency Gap?," *Journal of Economic Perspectives* 26, no. 1 (2012): 3–28, doi:10.1257/jep.26.1.3. A strong advocate of the gap is the consulting firm McKinsey & Company, for example, in *Unlocking Energy Efficiency in the U.S. Economy* (2009), www.mckinsey.com.

4. See Richard Thaler and Cass R. Sunstein, *Nudge: Improving Decisions about Health, Wealth, and Happiness* (London: Penguin, 2009).

Chapter 13. Green Political Theory

1. The material that follows relies on Paul Samuelson and William Nordhaus, *Economics*, 19th ed. (New York: McGraw-Hill, 2010), with a few changes for context.

2. A useful survey of the economics of liability law is in Robert D. Cooter, "Economic Theories of Legal Liability," *Journal of Economic Perspectives* 5, no. 3 (1991): 11–30, doi:10.1257/jep.5.3.11.

3. Mancur Olson, *The Rise and Decline of Nations: Economic Growth, Stagflation, and Social Rigidities* (New Haven, CT: Yale University Press, 2008).

Chapter 14. Green Politics in Practice

1. For data on battle deaths, see the Center for Systemic Peace, http://www .systemicpeace.org.

2. Kevin P. Gallagher and Throm C. Thacker, "Democracy, Income, and Environmental Quality" (PERI Working Papers, No. 164, 2008).

3. Madison is in Douglas A. Irwin, *Clashing over Commerce: A History of U.S. Trade Policy*, Markets and Governments in Economic History (Chicago: University of Chicago Press, 2017, Kindle), location 8425.

4. Data on tariff rates are from *Historical Statistics of the United States: Millennial Edition* (Cambridge: Cambridge University Press, 2006), table Ee430. Updated from the U.S. International Trade Commission.

5. Quoted in Irwin, *Clashing over Commerce*, location 8424–25.

6. Data on mortality from pollution are from Aaron J. Cohen et al., "Estimates and 25-Year Trends of the Global Burden of Disease Attributable to Ambient Air Pollution: An Analysis of Data from the Global Burden of Diseases Study 2015," *Lancet* 389, no. 10082 (2017): 1907–18, doi:10.1016/S0140-6736(17)30505-6.

Neal Fann, Charles M. Fulcher, and Kirk Baker, "The Recent and Future Health Burden of Air Pollution Apportioned Across U.S. Sectors," *Environmental Science and Technology* 47, no. 8 (2013): 3580–89, doi:10.1021/es304831q; E. W. Butt et al., "Global and Regional Trends in Particulate Air Pollution and Attributable Health Burden over the Past 50 Years," *Environmental Research Letters* 12, no. 10 (2017): 104017, doi:10.1088/1748-9326/aa87be.

7. Estimates of the marginal damages from sulfur emissions are for the year 2002, from Nicholas Z. Muller, Robert Mendelsohn, and William Nordhaus, "Environmental Accounting for Pollution in the United States Economy," *American Economic Review* 101, no. 5 (2011): 1649–75, doi:10.1257/aer.101.5.1649. The update to more recent years is from a personal communication from Nick Muller.

8. This estimate is based on a personal communication from Nicholas Z. Muller.

9. Céline Ramstein et al., *State and Trends of Carbon Pricing: 2019*, World Bank, 2019, doi:10.1596/978-1-4648-1435-8.

Chapter 15. The Green New Deal

1. The New Deal has been extensively studied by historians. A fine short history is contained in William E. Leuchtenburg, *Franklin D. Roosevelt and the New Deal* (New York: Harper, 1963). The authoritative history of the period is a two-volume work on Roosevelt by James MacGregor Burns.

2. Data on federal spending are from the Bureau of Economic Analysis, www.bea.gov, particularly table 3.9.5.

3. For example, see Robert J. Gordon and Robert Krenn, *The End of the Great Depression, 1939–41: Policy Contributions and Fiscal Multipliers*, National Bureau of Economic Research, no. w16380, 2010, doi:10.3386/w16380.

4. See, for example, Rolf Czeskleba-Dupont, Annette Grunwald, Frede Hvelplund, and Henrik Lund, *Europaeische Energiepolitik und Gruener New Deal: Vorschlaege zur Realisierung energiewirtschaftlicher Alternativen* [European energy policy and Green New Deal: Proposals for the realisation of energy-economic alternatives] (Berlin: Institut fuer Oekologische Wirtschaftsforschung [IOEW], 1994).

5. Thomas L. Friedman, "A Warning from the Garden," *New York Times*, January 19, 2007, and "The Power of Green," *New York Times*, April 15, 2007.

6. New Economics Foundation, *A Green New Deal*, 2008, https://neweconomics.org/2008/07/green-new-deal.

7. For the text of the House resolution, go to Recognizing the Duty of the Federal Government to Create a Green New Deal, H.R. 109, 116th Congr. (2019), https://www.congress.gov/bill/116th-congress/house-resolution/109/text?q=%7B%22search%22%3A%5B%22Green+New+Deal%22%5D%7D&r=1&s=2.

8. EIA, *Annual Energy Outlook*, eia.doe.gov.

9. Steven J. Davis et al., "Net-Zero Emissions Energy Systems," *Science* 360, no. 6396 (2018): eaas9793, doi:10.1126/science.aas9793.

Chapter 16. Profits in a Green Economy

1. From "Laudato Si': On Care for Our Common Home," encyclical letter, *Vatican Press*, w2.vatican.va.

2. Milton Friedman and Rose Friedman, *Free to Choose: A Personal Statement* (Boston: Houghton Mifflin Harcourt, 1990), 234, Kindle.

3. Steve Forbes, "Why the Left Should Love Big Profits," *Forbes*, May 7, 2014, https://www.forbes.com/sites/steveforbes/2014/05/07/profit-is-indispensable-for-prosperity/#4dc8455323b8.

4. Data on profits and capital are from the Bureau of Economic Analysis (BEA), particularly Sarah Osborne and Bonnie A. Retus, "Returns for Domestic Nonfinancial Business," *Survey of Current Business* 98, no. 12 (2018), www.bea.gov. The real yield on government bonds is the rate on the ten-year treasury minus the inflation rate. As a technical note, the returns data cover only domestic nonfinancial corporations. They exclude both financial profits and profits from foreign ownership. The more familiar data on earnings and stock prices, such as those on the Standard and Poor 500 stocks, are for publicly held companies, include financial companies and foreign earnings, and include elements of earnings (such as capital gains) that are excluded from the BEA estimates.

5. Nicholas Z. Muller, Robert Mendelsohn, and William Nordhaus, "Environmental Accounting for Pollution in the United States Economy," *American Economic Review* 101, no. 5 (2011): 1649–75, doi:10.1257/aer.101.5.1649.

Chapter 17. Green Taxes

1. George Washington, "Washington's Farewell Address" (speech), September 19, 1796, The Avalon Project, Yale Law School, transcript, https://avalon.law.yale.edu/18th_century/washing.asp.

2. Jimmy Carter, quoted in "Tax Reform: End the Disgrace," *New York Times*, September 6, 1977.

3. George H. W. Bush, "Acceptance Speech," delivered at the Republican National Convention, August 18, 1988, published December 4, 2018, by NBC News, https://www.nbcnews.com/video/1988-flashback-george-h-w-bush-says-read-my-lips-no-new-taxes-1388261955924.

4. Oliver Wendell Holmes, quoted in Compania De Tabacos v. Collector, 275 U.S. 87 (1927).

5. Markus Maibach, Christoph Schreyer, Daniel Sutter, H. P. van Essen, B. H. Boon, Richard Smokers, Arno Schroten, C. Doll, Barbara Pawlowska, and Monika Bak, *Handbook on Estimation of External Costs in the Transport Sector* (Holland: CE Delft, 2007); "Internalisation Measures and Policies for All External Costs of Transport (IMPACT)," *Handbook on Estimation of External Costs in the Transport Sector*, version 1.1 (Holland: CE Delft, 2008).

6. Estimates on environmental tax revenues are from the Organisation for Economic Co-operation and Development (OECD), *Towards Green Growth? Tracking Progress*, OECD Green Growth Studies (Paris: OECD, 2015), doi:10.1787/9789264234437-en.

7. The estimates on effective carbon tax rates are from Céline Ramstein et al., *State and Trends of Carbon Pricing: 2019*, World Bank, 2019, doi:10.1596/978-1-4648-1435-8.

8. Gilbert Metcalf, "A Distribution Analysis of Green Tax Reforms," *National Tax Journal* 52, no. 4 (December 1999): 655–82, doi:10.2307/41789423. His analysis considered primarily carbon taxes and taxes on conventional pollutants and totaled 10% of federal revenues.

Chapter 18. The Double Externality of Green Innovation

1. For award winners and several examples of successful research in Green chemistry, see the U.S. Environmental Protection Agency, "Green Chemistry Challenge Winners," https://www.epa.gov/greenchemistry/presidential-green-chemistry-challenge-winners.

2. This chapter draws on two of my earlier studies, *The Climate Casino: Risk, Uncertainty, and Economics for a Warming World* (New Haven, CT: Yale University Press, 2013) and "Designing a Friendly Space for Technological Change to Slow Global Warming," *Energy Economics* 33, no. 4 (2011): 665–73, doi:10.1016/j.eneco.2010.08.005.

3. The quotations on Green chemistry are from Paul T. Anastas and John C. Warner, *Green Chemistry: Theory and Practice* (Oxford: Oxford University Press, 1998); James Clark, Roger Sheldon, Colin Raston, Martyn Poliakoff, and Walter Leitner, "15 Years of Green Chemistry," *Green Chemistry* 16, no. 1 (2014): 18–23, doi:10.1039/C3GC90047A.

4. Paul Romer won the 2018 Nobel Prize in Economic Sciences for his studies in the economics of technology and new knowledge. A fine exposition of his work is found in Charles I. Jones, "Paul Romer: Ideas, Nonrivalry, and Endogenous Growth," *Scandinavian Journal of Economics* 121, no. 3 (2019): 859–83, doi:10.1111/sjoe.12370.

5. David I. Jeremy, "Damming the Flood: British Government Efforts to Check the Outflow of Technicians and Machinery, 1780–1843," *Business History Review* 51, no. 1 (Spring 1977): 1–34, doi:10.2307/3112919.

6. Geoffrey Blanford, James Merrick, Richard Richels, and Steven Rose, "Trade-Offs between Mitigation Costs and Temperature Change," *Climatic Change* 123 (2014): 527–41, doi:10.1007/s10584-013-0869-2.

7. Many of the potential new technologies along with strategies to promote them are contained in a special issue of *Energy Economics* 33, no. 4 (2011).

8. Some of these technologies may not be familiar. A description is contained in U.S. Energy Information Administration, "Electricity Explained: How Electricity Is Generated," https://www.eia.gov/energyexplained/electricity/how-electricity -is-generated.php.

Chapter 19. Individual Ethics in a Green World

1. Ethics is an enormous discipline. A short and enjoyable survey is Simon Blackburn, *Ethics: A Very Short Introduction* (Oxford: Oxford University Press, 2003). One of the foundational books in environmental ethics is Paul Taylor, *Respect for Nature: A Theory of Environmental Ethics* (Princeton, NJ: Princeton University Press, 1986).

2. See chapter 4's discussion. Also, as applying to the relations among countries, see John Rawls, "The Law of Peoples," *Critical Inquiry* 20, no. 1 (1993): 36–68.

3. For a discussion of climate ethics, see John Broome, *Climate Matters: Ethics in a Warming World* (New York: W. W. Norton, 2012). The discussion in this chapter draws on my review of Broome, "The Ethics of Efficient Markets and Commons Tragedies: A Review of John Broome's Climate Matters: Ethics in a Warming World," *Journal of Economic Literature* 52, no. 4 (2014): 1135–41, doi:10.1257 /jel.52.4.1135.

4. The environmental group the National Resources Defense Council (NRDC) has written wisely about the issues here. For useful advice on offsets, see "Should You Buy Carbon Offsets?," https://www.nrdc.org/stories/should-you-buy-carbon -offsets.

5. Ibid.

Chapter 20. Green Corporations and Social Responsibility

1. The three definitions of ESG are from Ronald Paul Hill, Thomas Ainscough, Todd Shank, and Daryl Manullang, "Corporate Social Responsibility and Socially Responsible Investing: A Global Perspective," *Journal of Business Ethics* 70, no. 2 (2007): 165–74; John L. Campbell, "Why Would Corporations Behave in Socially Responsible Ways? An Institutional Theory of Corporate Social Responsibility," *Academy of Management Review* 32, no. 3 (2007): 946–67, doi:10.5465 /amr.2007.25275684.

2. Milton Friedman, "The Social Responsibility of Business Is to Increase Its Profits," in *Ethical Theory and Business*, 8th ed., ed. Tom L. Beauchamp, Norman E. Bowie, and Denis G. Arnold (London: Pearson, 2009), 55.

3. Michael C. Jensen, "Value Maximization, Stakeholder Theory, and the Corporate Objective Function," *Business Ethics Quarterly* 12, no. 2 (2002): 235–56, doi:10.2307/3857812.

4. Ibid., 239.

5. A deep critique of market fundamentalism is contained in Amartya Sen, *On Ethics and Economics* (New York: Basil Blackwell, 1987). The quotation from Pope Francis is found in "Laudato Si': On Care for Our Common Home," encyclical letter, *Vatican Press*, w2.vatican.va. The Friedman quote is from Friedman, "Social Responsibility of Business," 55.

6. See Sen, *On Ethics and Economics*.

7. The court case cited is Sylvia Burwell, Secretary of Health and Human Services et al. v. Hobby Lobby Stores, Inc. et al. (2014), No. 13-354, June 30, 2014. Note that this is the argument used to allow corporations to use their funds for political purposes (which met widespread dissent), but it also makes the more general point that corporations do not have a narrow single purpose of shareholder value maximization.

8. William M. Landes and Richard A. Posner, "The Independent Judiciary in an Interest-Group Perspective," *Journal of Law and Economics* 18, no. 3 (1975): 875–901; William M. Landes, "Economic Analysis of Political Behavior," *Universities-National Bureau Conference Series* 29 (1975).

9. Christopher Stone, *Where the Law Ends: The Social Control of Corporate Behavior* (New York: Harper, 1975).

Chapter 21. Green Finance

1. USSIF Foundation, *Report on U.S. Sustainable, Responsible and Impact Investing Trends, 2018*, https://www.ussif.org/currentandpast.

2. TIAA-CREF, "Responsible Investing and Corporate Governance: Lessons Learned for Shareholders from the Crises of the Last Decade" (policy brief), published March 2010, https://www.tiaainstitute.org/sites/default/files/presentations/2017-02/pb_responsibleinvesting0310a.pdf.

3. California Public Employees' Retirement System, "CalPERS Beliefs: Thought Leadership for Generations to Come" (report), published June 2014, https://www.calpers.ca.gov/docs/board-agendas/201501/full/day1/item01-04-01.pdf.

4. The Yale report on ethical investment is John G. Simon, Charles W. Powers, and Jon P. Gunnemann, *The Ethical Investor: Universities and Corporate Responsibility* (New Haven, CT: Yale University Press, 1972), http://hdl.handle.net/10822/764056.

5. For the market, I use the Vanguard Total Stock Market Index Fund (VTSMX), which is a value-weighted fund of the largest thirty-six hundred U.S. public companies.

6. Summary scores for ESG for companies and industries can be found at CSRHub Sustainability Management Tools, https://www.csrhub.com/CSR_and_sustainability_information.

7. The calculations in the text rely on standard financial theory. Returns are assumed to have two risk components: a market component and an idiosyncratic component. An investor can reduce portfolio risk by diversifying, which allows the idiosyncratic parts to be reduced in importance. Table 21-1 assumes that each

company has an expected 6% real annual return and that half of each company's risk is market and half is idiosyncratic. Green portfolios have higher risk because they have less diversification. To compare different investments, I adjusted the portfolios to have the same risk by substituting safe bonds for risky stocks in the riskier portfolios. The table therefore shows the return penalty of different portfolios that have the same risk. The fundamentals of modern portfolio theory can be found in the enormously informative and entertaining book, Burton Malkiel, *A Random Walk Down Wall Street*, 11th ed. (New York: W. W. Norton, 2016).

8. The ESG average takes the expense ratio from Charles Schwab, *Socially Conscious Funds List, First Quarter 2020*, for U.S. equity funds, www.schwab.com.

Chapter 22. Green Planet?

1. This chapter builds on my book *The Climate Casino: Risk, Uncertainty, and Economics for a Warming World* (New Haven, CT: Yale University Press, 2013).

2. An extraordinary book on environmental treaties is Scott Barrett, *Environment and Statecraft: The Strategy of Environmental Treaty-Making* (Oxford: Oxford University Press, 2003).

3. There are many places to turn for a more complete discussion. An excellent textbook on climate is William F. Ruddiman, *Earth's Climate: Past and Future*, 3rd ed. (New York: W. H. Freedman, 2014).

4. Massachusetts v. EPA, 549 U.S. 497 (2007), https://www.supremecourt .gov/opinions/06pdf/05-1120.pdf.

5. The estimates here are from my Nobel lecture, William Nordhaus, "Climate Change: The Ultimate Challenge for Economics," *American Economic Review* 109, no. 6 (2019): 1991–2014, doi:10.1257/aer.109.6.1991.

6. These findings have been central to climate-change economics since its earliest days. A summary of the basic modeling and findings is contained in Nordhaus, "Climate Change," 1991–2014, doi:10.1257/aer.109.6.1991, with references to the literature contained there. Several chapters of the assessment reports of the Intergovernmental Panel on Climate Change (IPCC) have explored these points in depth. See www.ipcc.org for recent reports.

Chapter 23. Climate Compacts to Protect the Planet

1. The DICE (dynamic integrated model of climate and the economy) model is a computerized mathematical set of equations that represents the key components of the economy and the earth system. It can be used to project emissions and climate change and to test policies. A description is contained in William D. Nordhaus, *Climate Casino: Risk, Uncertainty, and Economics for a Warming World* (New Haven, CT: Yale University Press, 2013).

2. See, for example, my presidential address to the American Economic Association, "Climate Clubs: Overcoming Free-Riding in International Climate Policy," *American Economic Review* 105, no. 4 (2015): 1339–70, doi:10.1257/aer.15000001. A

nontechnical version of this is William D. Nordhaus, "Climate Clubs to Overcome Free-Riding," *Issues in Science and Technology* 31, no. 4 (2015): 27–34.

3. See Martin Weitzman, "Voting on Prices vs. Voting on Quantities in a World Climate Assembly," *Research in Economics* 71, no. 2 (2017): 199–211, doi:10.1016 /j.rie.2016.10.004.

Chapter 24. Skeptics of Green

1. John Aloysius Farrell, "Koch's Web of Influence," Center for Public Integrity, 2011, accessed May 19, 2014, https://www.publicintegrity.org/2011/04/06/3936 /kochs-web-influence.

2. https://www.greenpeace.org/usa/global-warming/climate-deniers/koch -industries/.

3. The passage from Kirkpatrick Sale is quoted in John Zerzan, ed., *Against Civilization: Readings and Reflections* (Eugene: Uncivilized Books, 1999).

4. The discussion and quotations are from Milton Friedman, *Capitalism and Freedom: Fortieth Anniversary Edition* (Chicago: University of Chicago Press, 2009), Kindle; and Milton and Rose Friedman, *Free to Choose: A Personal Statement* (Boston: Houghton Mifflin Harcourt, 1990), Kindle.

5. Milton Friedman and Rose D. Friedman, *Free to Choose: A Personal Statement* (New York: Harcourt Brace Jovanovich, 1980), 218.

6. Data on historical emissions of sulfur dioxide are from Sharon V. Nizich, David Misenheimer, Thomas Pierce, Anne Pope, and Patty Carlson, *National Air Pollutant Emission Trends, 1900–1995*, EPA-454/R-96-007 (Washington, DC: U.S. Environmental Protection Agency, Office of Air Quality, 1996), with updates from the EPA. GDP data are from the U.S. Bureau of Economic Analysis, https://www .bea.gov/data/gdp, with early estimates from private scholars.

7. David Anthoff and Robert Hahn, "Government Failure and Market Failure: On the Inefficiency of Environmental and Energy Policy," *Oxford Review of Economic Policy* 26, no. 2 (2010): 197–224, doi:10.1093/oxrep/grq004.

8. William D. Nordhaus, *The Climate Casino: Risk, Uncertainty, and Economics for a Warming World* (New Haven, CT: Yale University Press, 2013).

9. Ronald Reagan, "Remarks on Signing the Annual Report of the Council on Environmental Quality" (speech), July 11, 1984, The Ronald Reagan Presidential Library and Museum, transcript, https://www.reaganlibrary.gov/archives/speech/ remarks-signing-annual-report-council-environmental-quality.

10. Estimates on pollution control expenditures are from the Bureau of Economic Analysis, *Survey of Current Business* (Washington, DC: Bureau of Economic Analysis, 1996).

Chapter 25. A Tour of the Spirit of Green

1. Barbara Tuchman, *The March of Folly: From Troy to Vietnam* (New York: Knopf, 1984), 7.

INDEX

accounting, national, 83–94
    in Biosphere 2 and U.S. compared, 107–108, *108*
    climate change corrections in, 89–91, *90*
    externalities in, 85–93, 316, 327n3
    Green, 83–94, 316
    gross domestic product in. *See* gross domestic product
    pollution corrections in, 86–87, 89, 92–93
    subsoil assets in, 91–92, 327n6
    Weitzman approach to, 85–87
acid rain, 163
activists in Green movement, 300–301
advanced combined cycle, 217
agriculture, climate change affecting, 274
air pollution. *See* pollution
airport congestion, 25, 37, 198–199
alcohol
    excluded in ethical investments, 255, 259
    taxes on, 199, *200,* 318
Amazon.com, Inc., 182, 242, *260,* 261
anarcho-primitivism, 300
animals
    fairness to, 65–69
    rights of, 10, 66, 300
Antarctica, 103, 109, 272, *272,* 274
Anthoff, David, 309
anthropocentric approach, 9, 10–12, 300
apartheid, 258
Apple, Inc., 183, *260*
appliances, energy use of
    discounting of, 131, 132
    informational deficiencies on, 31
    principal-agent problem in, 55
    regulatory approach to, 140
appropriability of innovations, 207, 209, 210–211, *211*

architecture, 2–3, 55, 140
asbestos, 251
asteroids, 104, 111, 113
atmosphere
    as common property resource, 43
    historical carbon dioxide concentration in, 271, *272*
atomic weapons, 113
auctions
    of scarce public resources, 198–199, 202
    of sulfur dioxide emissions permits, 196–197
automobiles, 197–198
    costs of air pollution from, 22
    culture of, 38
    electric, 38
    federal regulations on, 51–52, 143, 144
    and gasoline tax, 64, 198, 201, 203, 318
    and highway congestion. *See* highway congestion
    innovation of, 15, 305, 319
    optimal pollution from, 47–49

Barrett, Scott, 268
Bator, Francis, 34
Becker, Gary, 312
Becquerel, Edmond, 219, 220
behavioral anomalies, 23, 129–140, 317–318
    discounting in, 75, 130–132, 317
    first-cost bias in, 130, 132–133, 136, 138, 140, 317
    inefficiencies in, 28, 129, 130
    life-cycle analysis in, 136–139, 317
    noneconomic preferences in, 135
    regulatory approaches to, 139–140
    solutions to, 136–140
    sources of, 133–135
Bell Telephone Labs, 220
Bento, Antonio, 64

A NOTE ON THE TYPE

This book has been composed in Adobe Text and Gotham.
Adobe Text, designed by Robert Slimbach for Adobe,
bridges the gap between fifteenth- and sixteenth-century
calligraphic and eighteenth-century Modern styles.
Gotham, inspired by New York street signs, was designed
by Tobias Frere-Jones for Hoefler & Co.